Palgrave Studies in Media and Environmental Communication

Series Editors

Anders Hansen
Department of Media and Communication
University of Leicester
Leicester, United Kingdom

Steve Depoe
McMicken College of Arts & Sciences
University of Cincinnati
Cincinnati, Ohio, USA

Advisory Board:

Stuart Allan, Cardiff University, UK
Alison Anderson, Plymouth University, UK
Anabela Carvalho, Universidade do Minho, Portugal
Robert Cox, The University of North Carolina
at Chapel Hill, USA
Geoffrey Craig, University of Kent, UK
Julie Doyle, University of Brighton, UK
Shiv Ganesh, Massey University, New Zealand
Libby Lester, University of Tasmania, Australia
Laura Lindenfeld, University of Maine, USA
Pieter Maeseele, University of Antwerp, Belgium
Chris Russill, Carleton University, Canada
Joe Smith, The Open University, UK

Aim of the Series
Drawing on both leading and emerging scholars of environmental communication, the Palgrave Studies in Media and Environmental Communication Series features books on the key roles of media and communication processes in relation to a broad range of global as well as national/local environmental issues, crises and disasters. Characteristic of the cross-disciplinary nature of environmental communication, the books showcase a broad variety of theories, methods and perspectives for the study of media and communication processes regarding the environment. Common to these is the endeavour to describe, analyse, understand and explain the centrality of media and communication processes to public and political action on the environment.

More information about this series at
http://www.springer.com/series/14612

Susanna Priest

Communicating Climate Change

The Path Forward

With contributions by
Neil Stenhouse and Jessica Thompson

Susanna Priest
Camano Island, Washington, USA

Palgrave Studies in Media and Environmental Communication
ISBN 978-1-137-58578-3   ISBN 978-1-137-58579-0   (eBook)
DOI 10.1057/978-1-137-58579-0

Library of Congress Control Number: 2016957350

© The Editor(s) (if applicable) and The Author(s) 2016
The author(s) has/have asserted their right(s) to be identified as the author(s) of this work in accordance with the Copyright, Designs and Patents Act 1988.
This work is subject to copyright. All rights are solely and exclusively licensed by the Publisher, whether the whole or part of the material is concerned, specifically the rights of translation, reprinting, reuse of illustrations, recitation, broadcasting, reproduction on microfilms or in any other physical way, and transmission or information storage and retrieval, electronic adaptation, computer software, or by similar or dissimilar methodology now known or hereafter developed.
The use of general descriptive names, registered names, trademarks, service marks, etc. in this publication does not imply, even in the absence of a specific statement, that such names are exempt from the relevant protective laws and regulations and therefore free for general use.
The publisher, the authors and the editors are safe to assume that the advice and information in this book are believed to be true and accurate at the date of publication. Neither the publisher nor the authors or the editors give a warranty, express or implied, with respect to the material contained herein or for any errors or omissions that may have been made.

Cover illustration: © jbdodane / Alamy Stock Photo

Printed on acid-free paper

This Palgrave Macmillan imprint is published by Springer Nature
The registered company is Macmillan Publishers Ltd. London
The registered company address is: The Campus, 4 Crinan Street, London, N1 9XW, United Kingdom

QC
902.9
.P75
2016

Preface

A few months into our work on this book, I attended a science café meeting in a bookstore in downtown Olympia, Washington. For those who haven't come across these, science cafés are local gatherings that give small groups of citizens a chance to meet with scientists and other community members interested in science and to discuss related issues in a face-to-face setting. These events are both recreational and educational; they have become an increasingly popular strategy for reducing the perceived gap between science and society in many places in the world. Few people know individual scientists on a personal level, so the thinking goes, unless they themselves also work in science, so events like these might help science (and scientists) seem more a natural part of life and less foreign or distant. Science cafés were not originally designed as community lectures, but in many areas they have become so popular that only a lecture format seems practical.

Olympia, my home when I first started this project and the state capital of Washington, has always struck me as a politically vibrant and progressive community—the type of place where artists, musicians, and environmentalists feel as much at home as do the state government bureaucrats and elected representatives who also live there. My expectation that an Olympia science café audience would be fairly knowledgeable about climate change activism seemed a reasonable one. The typical speaker at one of these events (here in Olympia or elsewhere) is, of course, an actual scientist. However, this particular event—while organized by scientists—featured a photographer's work on polar bears as a stimulus for discussion of climate change. The small event space was packed with what looked like

50 or so people. The presentation was well done, if somewhat predictable given the ubiquity of polar bear images as iconic representations of the dire consequences of climate change—and of habitat destruction more generally, in this case the destruction of the sea ice that is the bears' natural home and their only refuge on ocean-going hunting expeditions. And the audience was clearly both interested and sympathetic.

Yet in the discussion period, when someone in the audience asked the speaker what organization working on this issue he felt most deserved their support, neither the photographer nor the scientist organizers nor anyone else present had an immediate answer. Several in the audience looked up expectantly. One had the sense that here was a well-meaning (and probably well-educated) audience who sincerely wanted to help, but they didn't know how. After a very long pause, the photographer suggested a well-known wildlife organization concerned with protecting endangered species. Fair enough, but this was *not* an organization whose mission is focused primarily on responding to climate change. Rather, their interest is more narrowly centered on specific species and their vanishing habitats. The work that this and similar organizations do is undoubtedly very important. But as vitally important as the wildlife habitat destruction issues associated with climate change might be, they are only indicators of the much larger problem that must be solved. The global-scale energy policy and lifestyle adjustments that are essential to addressing the underlying problem of climate change will be major, and they must address problems that go far beyond dangers to the ice-dependent bears. Without addressing the problem of climate change and its anthropogenic foundations, the only real ways to help these polar bears would appear to be to relocate them—or to put them all in zoos. Yet not one of the people in the room had another immediate suggestion.

After another long pause, someone else in the audience suggested another organization, a newer one with a strong online presence and a mission specifically focused on reducing carbon emissions worldwide—a cyber organization with a less generally recognizable name (though an increasingly familiar one to climate advocates). This was the only climate-specific activist organization anyone in the group managed to identify. Most people seemed not to have heard of it, and the suggestion didn't seem to spur much response, either. No pens came out of pockets, no one asked the speaker to "please repeat that", no murmur of relief spread through the crowd—and no one suggested another name. People started shuffling their chairs and belongings; it was time to get home. Would a more recognizable, more locally based program have met a different reception? Of course, I can't say, but the event set me thinking.

I had already been wondering for quite some time about the apparent absence of truly local grassroots organizations—or even national ones—centrally concerned with climate change issues. The café experience confirmed my impression that such groups can be hard to identify. If they are not visible here and are unfamiliar to this group of demonstrably concerned citizens, then where are they? And why aren't there more of them? As a social scientist and a communication scholar, this gap struck me as both very important and somewhat puzzling. What do we know, as social scientists, about why this is the case and how to overcome it? How can we motivate individuals to change their lives and influence others, to mitigate climate change and adapt to it, if we are starting with so little organizational infrastructure to support this?

Around this same time, the death toll from Typhoon Haiyan in the Phillipines was being reported as 3982 and climbing (McClam 2013). This was already many more than the roughly 3000 who are known to have perished in the United States as a result of the September 11, 2001, terrorist attacks, which as we all know changed U.S. foreign policy forever. And the Haiyan death toll kept climbing, reaching over 6000 by December, plus 1800 still missing, 27,000 injured, and nearly 4 million people forced to leave their homes ("Typhoon Haiyan death toll" 2013). Of course, no one can say that climate change is responsible for the severity of any one storm, yet we also know we are in for much more of the same if the rate of warming is not slowed; this paradox is one of the great challenges of communicating climate change.

Meanwhile, yet another round of United Nations talks on climate change, this one in Warsaw, was reported to be faltering; the *Washington Post* reported these reasons in an online story:

> Poor countries argue that richer countries are responsible for most of the carbon-dioxide *already* in the atmosphere, so they should pay for the damage caused by global warming. The rich countries, for their part, point out that you also have to look at future emissions when divvying up blame—which puts the spotlight on fast-growing nations like China and India (Plumer 2013).

But hadn't we heard these arguments before? Is anything the least bit "new" about this news? The world seems to have become locked in a stalemate at the level of global politics—and neither rich nor poor countries are likely winners in this battle. We are all losers. What will it take for the world to move ahead? I looked forward to the 2015 Paris climate talks that were soon to take place. These were more successful, to be sure—some would say astonishingly so—but their goals are far from realized, and realizing those ambitious goals may not even go far enough.

This book is not about climate change as a scientific phenomenon; those facts are well-known and already available in hundreds if not thousands of books, reports, and scientific articles. Whatever the true proportion of climate "skeptics" who are motivated by political ideology versus simple ignorance (or some combination thereof), this book is not particularly for them, either. This book is for the clear majority who believe that climate change is a problem that we must do something about, and especially for those who should be taking leadership roles in forging solutions: those teachers, fellow scholars, politicians, agency officials, science museum staff, scientists, religious leaders, naturalists and park rangers, conservationists, environmentalists, and ordinary good citizens who seek a path forward but do not necessarily know how to proceed, either individually or collectively. In particular, this book is for communication scholars who want to know how their own work might matter to finding new solutions. We would hardly pretend to have all the answers either, but as communication researchers ourselves, we do know that there is a lot of social science research that can help us in this quest.

And a quest it certainly is. A certain level of zealousness may be required to overcome decades of stagnation and political gridlock, as well as the lack of an organizational infrastructure at many levels on which to build, and to compete with an understandable, even rational, popular obsession with global economic issues, as well as (in the United States at present) a much less understandable and far less rational battle over healthcare reform. The latter is actually a battle over ideological supremacy, not factual reality. In the same way, climate change should not involve a battle over the facts at all—which, science shows us, are very clear.

Toward the end of his life, astronomer, author, and well-known science communicator Carl Sagan published a book called *The Demon-Haunted World: Science as a Candle in the Dark* (1996), taking for his title a vivid and rather stark metaphor about the battle between science and superstition. I have always been optimistic in this area, believing that people are naturally both smart and rational, and that given broad access to both education and information, they will make good decisions. In other words,

I believe in democracy. And my first academic degree was in anthropology, which taught me that if people believe in demons, this is not necessarily best understood as simple irrationality; rather, to understand them, we must imagine a world in which it *makes sense* that demons exist.

So why does it seem to *make sense* to so many of us, despite all the evidence to the contrary, that climate change could not possibly be happening? This book will consider that question, but then move on to consider what communication research and other social science has to say about both persuasion and collective action that may be of use in the quest to get beyond first instincts in this matter—as well as to *make change happen*. It will also call on researchers to refocus their efforts on this question and in some cases to follow new paths. We should remember that as the majority of us move on to develop new strategies and new solutions, as well as implement the ones we already know about, the remaining skeptics are likely to get on board—and even if they do not, it is past time for the rest of us to move forward.

We write this book primarily from a U.S. perspective, not because we particularly blame the United States alone for the current situation, or because the United States is such a major contributor to the global production of carbon dioxide and therefore to anthropogenic climate change, or even because the United States is such an important world leader, but primarily because this is the cultural context we know best. Even so, we hope and believe that the lessons learned from the research of the international community of scholars that we report on in this book will be applicable in other contexts and otherwise prove useful to those elsewhere. For those readers, it may be useful to think of this book as a case study of one of the countries near the top of the list of contributors to climate change.

Susanna Priest
Camano Island, WA, USA

References

McClam, Erin. 2013. 'Unlikely We'll Ever Know': A Grim, Chaotic Count After Phillipines Typhoon. *NBC News*. http://worldnews.nbcnews.com/_news/2013/11/18/21523361-unlikely-well-ever-know-a-grim-chaotic-count-after-philippines-typhoon?lite

Plumer, Brad. 2013. Why the U.N. Climate Talks Keep Breaking Down, in Five Simple Charts. *Washington Post*. http://www.washingtonpost.com/blogs/wonkblog/wp/2013/11/20/why-the-u-n-climate-talks-keep-breaking-down-in-charts/

Sagan, Carl, and Ann Druyan. 1996. *The Demon-Haunted World: Science as a Candle in the Dark*. New York: Random House.

Typhoon Haiyan Death Toll Tops 6000 in the Phillipines. (2013, December 13). *CNN*. http://www.cnn.com/2013/12/13/world/asia/philippines-typhoon-haiyan/

About this Book

This book is not designed to convince its readers that climate change exists or that it is caused primarily by human activities. Rather, it is for those who are already convinced of both of these points—and who wish to persuade others to take seriously the relevant evidence and to help create new paths to action. Many readers of this book will be people who do research on communication processes and effects. The book is not *only* for them, however. It is also for many others—including scientists themselves—who want to be better science communicators or who already are professional science communicators (such as science journalists, outreach specialists, museologists, climate advocates, or alternative energy specialists), who want to engage others on the topic of climate, and who are attracted to taking a research-based approach about what kinds of communication will work best.

Science communication research, which may once have seemed like an obscure specialty, has exploded in recent years. Communication research may be our newest social science, but *science* communication research is the newest subdiscipline within it. Even so, we do not have all the answers, of course, and in putting together this book we did not aspire to include all of the literature available, which would be a daunting task indeed. Excellent work is being done by scholars around the world, and we could not mention them all. Rather, this book identifies some of the central themes and trends in the existing literature that seem most likely to inspire both scholars and practitioners. It also attempts to sketch some new directions that have not yet been researched as thoroughly, but that we believe have particular promise. We intend its message to be an optimistic one: Climate change is real, and we can come together to address it.

I thank the University of Washington's Department of Communication in Seattle for providing me with office space and collegial support during the initial conceptualization and early drafting of this project. I also thank my two contributors, Neil Stenhouse and Jessica Thompson, for reading some of the draft material as well as very generously sharing portions of their own work for inclusion here.

Contents

1 **The Communication Challenge of Our Century**   1
    Deficit Versus Dialogue   5
    "Messaging" and the Unit of Analysis Problem   9
    Strategic versus Democratic Goals   15
    The Path Forward   18

2 **What's the Rush? Reacting to a Slow-Moving Disaster**   23
    Understanding "Skepticism" and Inaction   25
    Change in the Information System   28
    Communication Research: Contributions and Limitations   30
    Distinguishing the Social from the Individual   32
    Journalistic "Objectivity" and the Cultivation of Uncertainty   34
    Scientists as Key Communicators; Non-Scientists as Audience   37
    Can the Climate of Opinion Actually Be Changed?   39

3 **Talking Climate: Understanding and Engaging Publics**   43
    Public Understanding of Climate Science   47
    Ideological Commitments Matter: Politics, Worldviews,
        and Values   50
    Trust and Efficacy   54
    What Have We Learned?   58

4	**The Evolving Social Ecology of Science Communication**	65
	Organizations and Institutions as Agenda-Builders	66
	Professional Associations for Journalists and Scientists	70
	A Word about the Nature of Social Norms	73
	Evolving Journalistic Ethics	75
	A New Ethical Landscape for Scientists?	80
5	**Science Communication: New Frontiers**	89
	The New Knowledge Brokers: New Media, New Actors	95
	Beyond the Ivory Tower	102
	New Audiences: Active Information Seekers	107
6	**Critical Science Literacy: Making Sense of Science**	115
	The Social Side of Science—And Why It Matters	119
	Whose Fault Is All This Confusion?	124
	Redefining Science Literacy	128
	Going Forward	132
7	**Ingredients of a Successful Climate Movement**	137
	Gaining Attention for Climate	139
	What Scholarship Tells Us About Social Movements	143
	Lessons and Opportunities for Communication	152
8	**The Path Forward: Making Change Happen**	161
	Keep Talking! Interpersonal Strategies Matter	165
	Focus on the Collective: A Renewed Research Paradigm	166
	Action Orientation: Climate as a Social Justice Issue	168
	Push Out Solutions, Not Just Problems	170
	Index	173

CHAPTER 1

The Communication Challenge of Our Century

Climate change is already present in the world, and much of it is caused by human beings. The purpose of this book is not to further establish or illustrate these facts, already supported by a strong scientific consensus, yet again. Rather, the purpose of this book is to examine the potential of improved, to some extent redirected, research-based communication efforts to contribute to the development of both individual action and collective societal will. We need to press for new policy solutions at the local, national, and global levels. We believe that a broad-based approach is essential to fully enable mitigation and adaptation with respect to our changing climate, as well as to identify some of the important research gaps and possibilities presented by the present communication environment. In order to move toward accomplishing these goals, the book considers and also critiques some of the streams of communication research that have generally dominated approaches to the issue of climate, pointing out a number of lessons, challenges, and opportunities. For those of us sincerely concerned about making a positive difference, communication research can help (although of course it cannot illuminate everything). Some paths and some concepts emerging from communication and other social science research—much of this work going back many decades— have proven very likely to be helpful at the present juncture. However, myths about the nature of public opinion on science-related issues persist, and some of the available research paths remain seemingly neglected or forgotten, underexplored and underexploited in the fight against the present climate stalemate.

© The Author(s) 2016
S. Priest, *Communicating Climate Change*,
Palgrave Studies in Media and Environmental Communication,
DOI 10.1057/978-1-137-58579-0_1

This book proposes that the communication research community and others interested in addressing the question of climate change, including scientists and practicing communicators, need to rethink some of their assumptions and approaches—while redoubling their already extensive efforts. In so doing, the book tries to chart a few new directions for the field of communication studies (and its newest subdiscipline, science communication studies), urging something of a return to the field's roots in *social* science and a departure from over-dependence on both thinking and methodology drawn from research focused on the psychology of individuals rather than groups. We fully acknowledge that much of that work remains important and can guide some forms of practice very effectively, and we briefly survey it here (especially in Chap. 3). However, it is a key premise of this book that public opinion and societal will are, at root, collective phenomena rather than simply characteristics of individuals. Better understanding these phenomena and what it means to consider them at the collective level can also provide us with a better understanding of what it means to be human, to be an individual who may in some cases live and work almost entirely alone—an option supported by contemporary technology—but who is inevitably tied by a thousand cords to the collective whole in which political, economic, and cultural dynamics that far transcend the individual play out. While often insightful, individual-level approaches to persuasion represent only part of the much bigger story. What we need in order to move forward is a multi-pronged approach in both research and practice.

In the United States, in late 2015, some important signs of positive change began to appear on the horizon—as well as some other important signs that appeared headed in the opposite direction. U.S. President Barack Obama had finally begun to speak openly about this issue and to raise it with other international leaders after years of near silence. Pope Francis also visited the Americas and came forward calling for action on climate, a hopeful development even for those of us who are not conventionally religious: The Catholic church claims well over a billion members worldwide, and the Pope has respect even beyond that group. Since public opinion is influenced by opinion visibility and the influence of that visibility on perceptions about what others think, these individuals and others like them—the type of person that social scientists have long called "opinion leaders"—play a vital role, even where it is primarily a symbolic one. Opinion leaders can exist on many levels—within a family, a neighborhood, a province or state, a professional or cultural group, or a nation.

Obama and Francis are truly global opinion leaders, in positions allowing them to exert influence around the world. To get the climate message across, though, we also need opinion leaders at all levels and in many walks of life who follow their lead and speak openly and with a sense of urgency about the existence and importance of climate change.

And public opinion, as measured by poll results, continues to evolve. While the United States has sometimes been compared unfavorably to Europe, where recognition of climate problems seemed to have emerged earlier and more strongly, the Yale Project on Climate Change Communication reports that as of 2014, 63 % of Americans believed that climate change ("global warming") was happening, and even though only 48 % believed that this is caused mainly by human activities, 77 % nevertheless supported research into renewable energy, 74 % supported regulating CO_2 as a pollutant, and 63 % supported strict controls on CO_2 emissions for coal-fired power plants (Howe et al. 2015). In other words, significant numbers of people support action on climate and energy even if they do not accept the idea of human causation of climate change and even—for some points—if they do not accept climate change at all. It is past time to stop focusing on the estimated 18 % (according to this same report) who simply do not believe in climate change and to find ways for the majority who do believe the science to move forward with solutions. In the United States, at 63 %, this group is over three and a half times as many people as remain "unbelievers".

Yet here is the less welcome news: The news coverage of the U.S. presidential candidates that was unfolding in 2015, accompanying the historic run-up to the party primaries preceding the November 2016 presidential election, very often featured Republican party (right of center) contenders initially in the forefront in the race for their party's nomination who are climate skeptics, including Donald Trump (who has called climate change a hoax invented by the Chinese to destroy the American economy; Desjardins and Boyd 2015) and Ben Carson (who has called the climate debate "irrelevant" because temperature change is cyclic; Desjardins 2015). That individuals with these views could rise to national prominence, despite these unscientific perspectives on climate, in a nation that considers itself one of the most powerful on earth is not encouraging. Both Democratic party (left of center) candidates, Hillary Clinton and Bernie Sanders, stated they accept the reality of climate change, although neither made it a particularly central campaign issue. However, the science of climate in itself should never have become a political weapon in the first place. And the peculiar dynamics of this particular election cycle illustrate how easily the perception of collective public opinion can come to diverge from the reality, given the attention and implied credibility bestowed on the Republican candidates by the news media—thereby lending legitimacy and power to their "skeptical" positions on climate.

This issue of distorted perception of public opinion is itself a challenge for defining a path forward on communicating climate change. Major funding from a variety of often untraceable counter-movement groups and organizations (Brulle 2012), as well as from the energy industry (Frumhoff and Oreskes 2015), promotes "denier" perspectives on the science of climate. This not only misrepresents the scientific consensus, it also contributes to the misperception that Americans more generally do not believe in climate change and do not want to do anything about it. The more denier voices are heard and seen, the more they will seem like majority opinion—even if they are not. The issue is not limited to the direct effects of expressing "denier" ideas, or defining them as "scientific". Public perception of the range and valence of public opinion is also important.

News media cannot resist covering the provocative statements of candidates like Trump and Carson or other public figures, even though they are not actually typical of American or even necessarily Republican views. The Yale project has also highlighted differences within the Republican party, and their figures show that liberal and moderate Republicans also believe climate change is occurring (at 68 % and 62 %, respectively; Howe et al. 2015). It can be extremely important for those in the majority to realize that they are in the majority, part of what this book will discuss in its upcoming chapters. The new message should not be "climate change exists" or even "climate change exists, and it is caused by humans", but "climate change believers are a strong majority, and they want action".

The present chapter is intended to provide us with a few specific starting points for considering these and other questions. We are concerned with the longer-term future, not just the immediate political context. We first need to grasp some of the recent trends and current thinking in science communication studies, which may be largely unfamiliar even to communication scholars who have not previously focused on science-related issues and therefore may be unfamiliar with these arguments, as well as to many natural scientists and communication practitioners. Science communication scholarship and practice have undergone major change over the past several decades. Science educators (including most practicing university-based scientists) are accustomed to focusing, naturally enough, on conveying accurate scientific information to their students. But for changing both individual and collective opinion, and for generating broader social change, that may not always be the most important thing—and it may not always be enough. To continue to move toward solutions, we also need to rethink a research paradigm that presently tilts sharply in the direction of studying individuals rather than groups. And we need to think more deeply about the broader purposes of communicating science, including climate science, as well as about how public opinion is actually formed.

Deficit Versus Dialogue

Both scholarly and applied approaches to science communication are sometimes described as having moved from "deficit" to "dialogue", a change that is in itself something of a social movement. Like most such catchy phrases that become popular, within as well as outside of academic circles, this shift means somewhat different things to different people. Changing attitudes toward science, or opinions on science-related issues, would certainly seem to require awareness of the science itself. No one who supports a science-based solution to something like climate change would really argue with this. However, at the same time, most people do not need or want to know *all* the science available, in order simply to make up their minds about an issue. They just need "enough" science—what is sometimes called "satisficing" or reaching some subjective internal threshold of perceived adequacy. How much information is enough is a matter of individual perception, in other words. And giving out information about science, while it can certainly help people reach that threshold, is not necessarily the best way to change people's attitudes or beliefs, especially once their minds have been made up. It is not enough to change minds (for climate, "deniers" simply do not believe the evidence, however much they are offered), and it is also not enough to motivate action, in many cases—even though it might help explain what kind of action will help and why.

To illustrate how this works, consider a different example. I can have an opinion about whether we should continue to invest in the space program without actually knowing how to build a rocket ship. I just need to know that the engineers know how to do this, so that the investment (not to mention astronauts' lives) will not be too easily wasted, and I also need to have a general sense of what society will gain from that investment. On these specific points, I usually need to trust the judgments of the scientific and engineering communities. Broader policy issues, on the other hand, are often largely either value judgments or strategic choices; whether investing in the space program is a better choice than investing in some other program is not entirely a "scientific" question. We hope that related public opinion will be scientifically informed, but it is not necessarily determined by science. Giving me more information (in this case, about the space program—or, say, about space science more generally) is unlikely to change my mind, unless perhaps it also excites and inspires me.

To consider another example, people generally support investments in medical research because they value good health and they can see an obvious benefit to themselves and others. This doesn't depend on their knowing everything that their doctor knows. Indeed, linking the general issue of climate change to a desire to protect our public health is a reasonable strategic approach, one some strategists have been pursuing and that even U.S. President Barrack Obama has occasionally mentioned (Subramanian 2013). Most people probably will not want all the details, however; it would ordinarily be enough simply to know that there is a link that makes sense in a general kind of way. For this, they need to trust the messenger as much as they do to understand the science.

Importantly, the same people who support health research or the space program may or may not see the benefit in supporting more basic research (say, in biology or astronomy) or environmental research (say, finding the best way to protect the habitat of an endangered species). The values that each of these activities appeals to are distinct. And when weighing benefits and costs, the gains involved are not always tangible economic ones. For example, for those who find natural environments inherently valuable, protecting habitat is inherently meaningful. And space initiatives clearly have symbolic value for the national pride of participating nations, as well as knowledge gains reflected in both basic science and in "spin-off" innovations.

"Pure" science, on the other hand, is sometimes likened to art as something society should value for its own sake, and different people will disagree on how important this is. In other words, reasonable people can vary in how much they value different societal investments, including investments in different forms of science—or in science generally—and those opinions do not depend solely on the known scientific facts. Neither do other science-related beliefs. It is unlikely that improved knowledge of science will dissuade people from rejecting evolution if their underlying worldview insists that God, not biology, is responsible. (There is no compelling evidence of a single, monolithic, "anti-science" perspective; each specific issue is different.) To persuade people to invest time and resources in mitigating climate change, we need to appeal to their beliefs and values, not just get them to accept the science.

The idea that providing an abundance of scientific facts will necessarily alter people's opinions in a direction more consistent with the opinions of the scientific community has generally proven false based on the empirical evidence. That is in large part why science communication scholars now refer to this older way of thinking as the "deficit model"—the belief that public opinion (and public relations) problems can be solved with improved dissemination of scientific information alone. For example, neither education nor knowledge is closely linked to opinions about biotechnology (Priest 2000). This is not to say that knowledge is unimportant (Sturgis and Allum 2004), but only that the relationships here are much more complex and nuanced than observers might initially suppose—and that people's opinions are based on many factors other than possession of the scientific facts, notably their values and beliefs. Unfortunately, the so-called "deficit model" that assumes people have a knowledge deficit relative to science and so teaching the science will "solve" the "problem" continues to be reinvented. Many people, including many natural scientists, intuitively assume that it is obvious that improving science literacy will get people on their side. However, this is just not necessarily the case.

Turning away from the "deficit model" toward a more "dialogic" or "public engagement" approach has become associated with a variety of different directions for both scholarship and practice in science communication. More emphasis is being placed on creating opportunities for two-way communication and discussion (that is, for dialogue), both through improving scientists' opportunities and motivations for engaging with the public (and with policy-formation processes) and through improving non-scientists' opportunities and motivations for engaging with science—and with scientists. Opportunities to discuss issues related to science and science policy help people to think more deeply about them as they make up their minds. They are a valuable aspect of democracy, which current thinking in science communication recognizes.

Such opportunities will not necessarily change opinions, however, or predictably move them closer to the opinions of the relevant scientific communities—any more than further understanding the science itself will necessarily change people's basic underlying beliefs. Nor will everyone be interested enough in science or science policy to participate in discussions of science-related issues, even given the opportunity. Many—possibly most—will continue to rely on guidance from trusted opinion leaders. This means climate communicators should give thought to which leaders might be influential with particular groups, just as advertisers so commonly rely on famous spokespersons.

Newer work (in both scholarship and programmatic practice) within science communication has turned to a variety of more-or-less novel public engagement models, involving everything from "citizen science" in which non-experts participate in gathering and analyzing actual scientific data, to community science festivals and "science cafés" in which people come together in informal settings to interact with scientists, to more highly organized opportunities for citizens to discuss science-related policy issues (sometimes taking the explicitly deliberative form called "consensus conferences") and a broad range of science center and science museum-based efforts that emphasize hands-on demonstration and interactive experience—often with new opportunities to both discuss and question—in preference to passive displays. This "new wave" of science communication activities generally redefines outreach as a process of giving citizens opportunities to share their opinions with the scientific community, not just the other way around.

This is a promising direction, given the shortcomings of the "deficit model", even though it is not a panacea for "problems with the public". For one thing, a lot of people who are attracted to "public engagement" activities about science are already very interested in science; a few may be critics, but many are likely to be boosters. Some academic "engagement" exercises are artificial events, ones that would not necessarily have taken place without being funded as a form of social experiment. Sometimes such experiments are criticized for not providing a ready means for the outcomes of discussions to be considered in the political process, which participants often seem to expect. Planned discussions seem particularly unlikely to change attitudes where people are already polarized. This was dramatically illustrated by the Obama-era "town hall" discussions of the need for healthcare reform in the United States, events largely (and unexpectedly) dominated by strident anti-reform voices.

Engagement events are not designed and should not be thought of as mechanisms for getting "the public" on the side of a particular policy solution to a problem. When governments turn to such events to resolve pre-existing science policy disputes, as in the United Kindom's "GM Nation" discussions of genetically modified crops (see Horlick-Jones et al. 2006), they also open themselves to criticism for attempting to engineer hoped-for outcomes. Consensus itself often remains elusive and perhaps should not even be expected as a consistent outcome of discussion.

But beyond these observations, the main limitation of adopting public engagement approaches as a primary solution to the climate change problem may simply be that climate change is something of a communication emergency. It is my personal belief that given reasonable opportunities for thoughtful consideration, such as by participating in debate and discussion, people will arrive at wise collective decisions—even on highly technical matters.[1] Even so, it is hard to point to stunningly successful examples where this has happened on a large scale, at least as a result of deliberately orchestrated debate. And making difficult choices about addressing climate change could take years or even decades of public discussion, when we simply do not have years and decades available. Climate change is happening now, and it is going to prove increasingly irreversible.

We certainly need to both disseminate more factual information about climate change (just as "deficit model" adherents thought, but without presuming this is the whole solution), as well as facilitate more public discussion about climate-related issues (just as the move toward more public engagement and two-way communication between scientists and others suggests). Yet these are only the beginning, and they may not move us ahead far or fast enough. Many communication researchers seek instead to identify the most persuasive type of message. Unfortunately this approach has its limits as well.

"Messaging" and the Unit of Analysis Problem

Communication research and study is often divided not only according to the type of subject being communicated (such as science, politics, or health) but also by scale or scope (interpersonal, organizational, mass, and international/intercultural communication), medium or genre (print versus broadcast versus "new media" contexts; formal speeches versus conversations versus group discussions; visual versus text-based messages), as well as professional purpose (advertising or public relations versus various forms of journalism, old and new). These sorts of divisions give us identity and help us organize our teaching and other work as scholars, but they may also make it harder to see a bigger picture in a situation where considering just one form of communication at a single level provides a singularly myopic view of a much more complex process. And add to this mix the many different methodological repertoires that are applied to this range of communication research subspecialties, both qualitative and quantitative, the result is a cacophony that is not easily harmonized. The academic field of communication studies must seem as strange to some observers as theoretical physics does to others—appearing as though something that should be simpler, in other words, has been made remarkably esoteric.

Our research as communication scholars routinely concerns audience (that is, consumer) reception, but we often tend to reduce this to research about the outcome to be expected when a single individual encounters a single particular message, such as an advertising or political campaign message (in mass communication research) or something that is said (in interpersonal communication). Arguably, it is simply easier to study individuals as individuals than to study groups. And it is definitely easier to study things that happen in the short term (such as the answers we get to survey questions or outcomes based on factors we can control in an experiment) than in the medium or long term (such as the dynamics underlying trends in media content or changes in the tenor of popular thinking). We can sometimes forget that people are always thinking and acting as members of groups—in a pluralistic society, often as members of many diverse, interconnected, and overlapping groups—and respond in ways that reflect their varied group identifications (and not just their political party affiliations; see Pearson and Schuldt 2015 for specific climate-related discussion of this question). Individuals are also acting as participants in a broader culture that crosses group lines and that itself changes over time.

The difference between studying the reactions or characteristics of individuals or the collective behavior of groups is what social scientists call a "unit of analysis" problem, and it is especially relevant to approaching the collective nature of public opinion phenomena. To consider a simple example, we commonly encounter unit of analysis problems in data analysis, whether qualitative or quantitative. If we study a focus group transcript (say, of a group of people discussing what can be done to limit climate change), should we analyze each spoken sentence separately? Or should we base our analysis on each longer comment, made in a single conversational turn? That might be more sensible, considering that making a single point might take more than one sentence. But maybe, if we are looking at a large set of such discussions, we should instead be looking at some of the characteristics of the entire discussion in each case—did different groups go in different directions? The correct choice depends on our research question; are we intending to study how people react to the topic—or how the group affects the individual? Similarly, if we study print media coverage (say, newspaper coverage of climate), should our unit consist of whole publications such as newspapers, or daily issues, specific articles, sources quoted, paragraphs, or statements? These are important research design considerations. Our number of cases, or "n", is different for each choice. So is the meaning of the results.

Consider also, as a simple example of why collectivities matter, the generally recognized fact that focus group composition can affect the behavior of group members in important ways. For example, for technical topics in particular, people who feel they know less about the topic or who have less formal education may be easily intimidated in a discussion including experts or those with graduate degrees. Often focus group researchers try to organize groups that do not invite this problem by putting people together who have similar levels of education. Now translate this complexity to all of human society—especially in the context of a highly pluralistic society such as the United States. We are constantly surrounded by information (whether direct or mediated) about what other people think, and we are constantly adjusting to that knowledge in forming and expressing our own opinions.

Going against the grain of groups in our expressed opinions can be suppressed if we imagine bad consequences, as Elisabeth Noelle-Neumann famously argued through her "spiral of silence" theory (1993). If the group is one we identify with, we may fear rejection or isolation. This does not mean that we never disagree with our group—far from it—but we do tend to monitor how and when and to whom we express our disagreements. If we think few others in our group (neighborhood, religious group, community, political party, workplace, extended family, circle of friends, and so on) believe what we believe, we may or not make a point of expressing those beliefs in the company of that group. Whether we speak or are silent can, in turn, contribute to false impressions of group opinion—and in some cases such impressions may lead us to miss important opportunities to influence others. In short, our beliefs about what others in the group likely think are very important. This is exactly why having visible opinion leaders validate particular ideas and perspectives is important, although unfortunately this can also work in reverse: Outspoken "skepticism" about an idea (such as climate change) can validate others' rejection of that idea.

Yet in studying public opinion formation, even though we know that what we believe that others are thinking can be very important to the processes involved, communication researchers often turn first to consideration of the individual rather than the group to which that individual belongs (or, more importantly, with which that individual identifies, their "reference group"). Our data are often taken from surveys or experiments that look at individuals' reactions in isolation—often by design, since we are regularly trying to isolate or predict the effects of particular messages. Running statistical analyses that look for group differences based on things like demographic categories is a common option and a reasonable one in many situations, but this does not entirely solve the problem—especially if we have not asked the right questions to get beyond a simple "check box" reflection of group membership. Things like religion, political affiliation, or regional and class identity have many shades of meaning and influence, especially in different combinations. The whole is greater—and much more complex—than the sum of the individual parts (or in this case, the individual people). These influences are not reducible to "checkbox" demographics. Which groups are important to which people acting (or speaking) in which context is not easily captured by this kind of data.

Even more important, however, other phenomena within society that are vitally important to addressing a problem like climate change cannot really be studied at the individual level at all. Action for social change involves collective behavior, not just individual opinions or even individual action. Yes, individuals' opinions matter because they vote, they make decisions about their own energy use, they may sign petitions or participate in surveys or write letters and emails that have some broader influence—and they may remember to mention to their neighbors that they believe in climate change. But it is groups—from community non-profits to professional societies to political parties to corporations and whole industries—that most commonly advocate for change or resist it, and it is political institutions that collectively determine what regulations will be put in place and how rigorously they will actually be enforced. Individuals matter—especially in our individualistic society—but their influence is constrained by group processes. And individuals acting alone do not have the same social "weight" as groups acting together.

Studying group dynamics over time is often messier and may be perceived as less precise than studying individual-level phenomena such as how a particular individual may be influenced by a particular message at a single point in time. Further, many of the obvious practical applications of communication research—in political contexts, for example, or in product marketing or promoting healthier choices ("social marketing")—are concerned with individual-level behaviors (voting, purchasing, choosing).

We cannot successfully sell climate change action in quite the same way, since the end result needs to be a shift in the opinion landscape itself that enhances not just knowledge or belief on the individual level but perceived collective belief, collective efficacy, collective will, and ultimately collective action. Individuals acting purely as individuals are probably not going to be quite enough, even if one hundred percent of them were to believe in climate change.[2]

Mediated communication is important not only because it disseminates information but because it represents public opinion. It is often media accounts (including entertainment along with news) that give us our primary "window" on what other people think—whether opinion leaders, experts, or just ordinary people like ourselves, and whether the picture we see is accurate or distorted. Based on many thousands of studies, mass media do not often have the strong effects that the field once expected to find through its early research. But they do have effects. Regardless of what is said about an issue in the media, if nothing were said at all we might not even recognize that the issue exists, and the more it appears in the media, the more important we tend to believe it to be (McCombs and Shaw 1972; Iyengar and Kinder 1989). The media also tell us, directly and indirectly, about the range and diversity of opinion on an issue. They can and do suggest, sometimes in rather subtle ways, which modes of thinking are legitimate and which should be considered marginal or "fringe".

This is why the "false balance" coverage of climate change, which has portrayed climate change skepticism or denial as co-equal within the scientific community with climate change acceptance (see Boykoff 2011), has been such a particular problem. Likely as a result, many people remain confused about whether scientists disagree on climate change. According to the Howe et al. (2015) data cited above, just over one-third (34 %) of people in the United States still believed that scientists are in disagreement on this issue—almost twice as many as those who did not actually themselves believe that climate change is happening.

The dynamics of public opinion at a given moment are sometimes referred to as the "climate" of public opinion. Just like the weather (if not quite like the actual climate), public opinion sometimes seems to shift unpredictably and behave in unexpected ways, responding to forces we have not yet fully grasped and cannot always measure with confidence. What we perceive that different groups of people think, what they say, what the media report and who they quote, what and who we tend to believe (or discount), and our own individual senses of self-identity that provide guideposts with which to navigate this complexity all swirl around together in what is readily conceptualized as a huge cloud driven by strong winds. We do not know quite enough about how these winds work—but we do know that they can be powerful.

Media will continue to be important, and even though their actual effects are more nuanced than was once assumed, they are not limited to transmitting information. Few individuals are likely to rely on in-person engagement for most of their information about current issues in science, despite the importance of the trend toward innovative dialogic approaches that can facilitate two-way discussion and deeper engagement. But which media? With fluctuations in the global economy and the development of Internet-based communication alternatives of all kinds, traditional news media in many areas are undergoing restructuring, many newspapers have disappeared, and since many of the ones that are left are on tighter budgets, fewer experienced science or environmental journalists may be available to cover issues like climate. As it turns out, this matters as well.

We do not always know how these trends will influence the climate of public opinion, but its dynamics seem ever more complex. More information is available than ever before, but sharper skills are needed to sort out truth from fiction and journalism from public relations in an information environment with fewer reliable gatekeepers. Blogs, tweets, Facebook pages, videogames, ordinary websites, YouTube, and cell phone apps—to name just a few—provide new types of narratives about science and facilitate the formation of new virtual social networks and novel forms of social identity. How individual people seek and interpret information in this evolving environment of personalized media is the subject of ongoing research, but there are still many layers to unravel.

Later chapters will address most of these issues in more detail, but both mediated and interpersonal communication will continue to be important drivers of the public opinion climate for climate change, as for other public issues. However, in our research into the influence of the growing complexity of the information and communication landscape on individuals, we need to remember that analysis based on larger social units of all kinds should not be neglected. Climates of opinion are generated and have their influence at the collective level, not only at the individual level. Social movements are also collective phenomena—and climate change needs a bona fide social movement parallel to the civil rights movement. If this seems an odd analogy, consider for a moment the disproportionate impacts that a shifting climate will have—especially for millions upon millions of disempowered people living in environments that are already marginal—as deserts expand, lands dry out and heat up, coastal areas are further inundated, and storms grow ever more severe.

Strategic versus Democratic Goals

This discussion now brings us to the third foundational issue of particular relevance to this book, in addition to understanding the "turn to dialogic" in science communication and the unit-of-analysis problem. What actually is the purpose of science communication? Under this seemingly simple question lies quite a bit of complexity. If the goal is not simply to transmit information about science, is it to persuade people to agree with scientists? In recent years there has been much talk about "framing" science, which is generally understood as communicating about science in a way that will be persuasive in achieving a strategic communication goal.[3] Perhaps science should indeed be communicated in ways that people can better relate to their everyday lives. Perhaps scientists should be willing to participate—even to intervene—in policy making, which also requires persuasion. But should science be "spun" in ways designed to have a specific effect? Many people, including many scientists, are not entirely comfortable with that idea. And it can also backfire: People who sense they are being manipulated may react negatively to both message and messenger. Yet there is no doubt that many efforts to communicate about science—whether coming from small advocacy groups or large corporations, government agencies or private foundations, research universities or scientific societies—are intended to persuade. These may be intended to persuade with respect to their positions on specific controversies, or they may be intended to persuade in a more general way with respect to how people think about science. They may even be sincerely intended to persuade citizens of something "for their own good"—something they need to know for their own benefit. But they are definitely intended to persuade.

An alternative way to think about science communication and its goals is that it provides a means to improve democratic governance by making scientific thinking and scientific evidence transparently available to citizens as they seek to make up their minds about policy issues—and about their own lives (considering, for example, information about nutrition or about earthquake risk). We might still have different ideas about how this democratic goal might best be served. Should only "settled" science be disseminated, as is generally argued by those who disapprove of discussing or reporting (outside of scientific meetings) results that have not yet been peer reviewed? Or should science "in the making" also be available for all citizens to consider, evaluate, and discuss? Experiments with open journal review and the expanding science-related blogosphere would suggest this is already occurring. Pluralistic democracies are generally based on the principle of open dissemination of information and free speech. Should there be exceptions for science—even rather narrow ones, such as withholding the details of so-called "dual use" technologies that could be made use of as weapons? It seems reasonable to restrict some information, but lines like this are always difficult to draw. Regardless of the tactics involved, however, the goal of improving democracy is in principle quite different from the goal of persuasion. It is not always clear that the two are entirely compatible.

Generally speaking, in the world of professional communication, it is public relations specialists who are associated with strategic communication efforts, while journalism seeks to chart a lofty middle course that accommodates multiple points of view in the interests of maintaining a healthy marketplace of ideas, thus improving democracy. But this is a very blurry line. Journalism has been described as routinely (and largely uncritically) "selling" science (Nelkin 1995); it is also highly dependent on press releases and other "information subsidies" (Gandy 1982) disseminated by strategic communicators. Public information and outreach specialists who work for universities and government agencies may be motivated by a very sincere desire to disseminate neutral information designed to improve the lives of citizens, but they are also practicing a kind of public relations on behalf of their institutions by bolstering their reputations. This line has become even more blurred with the rise of the Internet as the central communication medium of modern times. As we will discuss further in a later chapter, it is increasingly up to the consumer of information to distinguish between good information and bad, a complex task that requires a new kind of science literacy (Priest 2013).

The classic role of the news media in modern Western democracies is to serve the interests of democracy by providing diverse points of view. We assume that citizens, as news consumers, will be the final arbiters of truth in this marketplace of ideas. Media discussion can also be said to serve as a stand-in for actual discussion in which it is not practical that everyone affected can participate. We cannot all participate in political debates, but we routinely do so vicariously, thanks in large part to cable television. Media also report back to us on discussions that take place in political contexts (e.g., in Congress or Parliament or other legislative bodies) where not everyone can attend, and on the speeches and other public statements of political candidates that not everyone would otherwise have the opportunity to hear. And they can report on the outcomes of various citizen deliberations, although this is rather less common.

Journalists are generally trained not to privilege one political or ideological view over another but to present "both" sides, as though there were always just two. But when it comes to science, another truth-seeking paradigm competes with this one. Inside the scientific community, the truth is generally decided by peer review, replication, and—eventually—scientific consensus. This process certainly has its own politics, but it is not quite the same as politics outside the scientific community. Journalists may need to leave aside the idea that there is always a "left" and a "right" view on everything. Instead, the evaluation of scientific legitimacy, consideration of the weight of evidence (Dunwoody 2005) and the degree of scientific consensus, and providing sensible context for interpreting uncertainties become much more important. Clearly these are vital for effective science reporting. Both journalists and citizens also need to better understand the social organization of science in order to interpret scientific information (see Chap. 6). It is unclear whether typical non-scientist news consumers understand the nature of scientific truth in the same way that scientists normally do.

Yet asking journalists to sort true from false science, especially when reporting science that is still "in the making", or science that is challenged, is asking a lot. It is at the heart of the scientific method that all that we know as truth today may be revealed as falsehood tomorrow. This may be partly why some scientists are so reluctant to state anything with complete certainty—or even to see their research results taken up in policy debates. But perhaps this needs to change. Expressions of uncertainty have undoubtedly taken their toll on public acceptance of climate science. And science that may need to be revised tomorrow still illuminates our best, most reasonable, choices today.

And as has already been noted, climate change may be seen as a communication emergency. We do not have the luxury of time on our side. Whether or not this means that, ethically, it is acceptable for persuasive communication to replace more democratic approaches remains an open and quite philosophical question, but in reality, many of the actors involved are pursuing strategic goals. This situation creates some very deep but largely unrecognized tension in the science communication world that affects decisions we must make about communicating climate change. But one conclusion stands out: It is well past time to shift the public debate from *whether* climate change exists to *how* we are going to respond to it, and to shift our communication research paradigms accordingly.

The Path Forward

These three themes—that science communication has moved from deficit to dialogue, yet both approaches remain relevant; that science communication research has quite often been too focused on the individual rather than the social; and that the tension between strategic and democratic goals for science communication runs deep—will recur throughout this book. In recent years, the U.S National Academy of Sciences focused attention on the "science" of science communication through its Sackler Colloquium series.[4] Given the complex dynamics within the field, as well as the limits of common research approaches, science communication practice will nevertheless remain both art and science. In addition, the research side of the field should be driven by theory, not just data, including theories of the social as well as the individual; its practitioners and scholars should think reflexively about the goals and ethical dimensions of their work; and they should remain attentive to the role of values and beliefs in human decision making, not just knowledge or emotional reactions. Climate change is an opportunity to consider all of these issues in more depth. We first take a closer look at the current situation.

The trajectory of this book is as follows. The next chapter (Chap. 2) considers in more detail some of the key reasons why it is so difficult for many people to grasp and accept that climate change exists and that it is largely caused by human beings. These range from the psychology of individuals to the nature of social influences, as well as contemporary practices and trends in national and global news systems. It also raises the questions of how communication scholarship can help, particularly if that scholarship takes account of social as well as individual-level phenomena. Chapter 3 then provides an overview of what we actually already know about public understanding and how it is influenced by communication, primarily at the level of the individual, emphasizing the need to understand that there are many "publics" rather than just one "general" public, and many levels and forms of understanding and engagement. We are making important gains in terms of public understanding and acceptance, but the politicized nature of the issue, the pluralistic nature of the United States and most other modern societies, and the role of diverse values and beliefs in people's thinking about the issue of climate remain challenges.

In Chap. 4, the discussion turns to influences that are most visible at the social level. Organizations and institutions, not just individuals, influence what problems we see as important and how they are defined and understood. Some key organizational players here are government agencies, scientific societies, nonprofit organizations or NGOs, and professional associations—including those for journalists. All organizations have their own internal expectations for behavior or norms; the professional norms of both journalists and of scientists are particularly relevant to this discussion. Chapter 5 extends this discussion by looking at the ways that our relevant expectations and practices are evolving: We are experimenting with new forms of public engagement, new media forms, and other new forms of science communication. The relationship between scientists and journalists, long a focus of science communication scholars, remains important in this context—but is itself evolving. Audiences that we once thought of as passive are now increasingly seen as active information seekers. Then Chap. 6 takes up the issue of what those seeking information about climate and other science need to bring to the table to make sense of it: a critical awareness of how science is done that allows them some basis for discriminating between credible and questionable scientific claims.

Part of the solution to all this complexity lies in creating a strong demand for the development of climate solutions. Understanding is important but action may be even more important. As Chap. 7 argues, media organizations tend to focus their attention on events, often dropping issue-oriented coverage during periods of relative quiescence. To maintain public attention and motivate change, a bona fide social movement is required. The chapter also reviews some well-established characteristics of movements for social change. And then Chap. 8 makes the case that communication scholarship could contribute more than it does by turning more of its own attention toward the factors that create and sustain collective action, rather than remaining focused too narrowly on factors that persuade individuals. This is the path forward for climate—and perhaps for other contemporary issues as well.

Notes

1. This view is philosophically consistent, at least in a general way, with John Dewey's classic work on the role of education in a democratic society (see, e.g., Dewey 1997), although the idea was not taken directly from his work.
2. While we do routinely incorporate consideration of perceptions of others' perceptions when applying some of the theories commonly used in communication research, including the Theory of Planned Behavior (Ajzen 2012) and work on Third-Person Effects (Davison 1983), the unit of

analysis is still the individual. Important group influences, including the dynamics of opinion climates and the role of broader cultural contexts, cannot generally be meaningfully conceptualized or adequately measured on an individual basis.
3. Framing, as a concept, has an exceptionally long and varied history; it has been used by cognitive psychologists, sociologists, political scientists, linguistics scholars, economists, media and communication scholars, and professional communicators in a variety of quite different ways. The present discussion is limited to one of the common ways the word is currently used in the field of science communication, which is to indicate a message that has been strategically constructed in order to persuade.
4. For details, see http://www.nasonline.org/programs/sackler-colloquia/completed_colloquia/science-communication.html?referrer=https://www.google.com/

References

Ajzen, I. 2012. The Theory of Planned Behavior. In *Handbook of Theories of Social Psychology*, eds. P. A. M. Lange, A. W. Kruglanski, and E. T. Higgins, vol. 1, 438–459. Sage.

Boykoff, M. 2011. *Who Speaks for the Climate? Making Sense of Media Coverage of Climate Change*. Cambridge University Press.

Brulle, R. 2012. Institutionalizing Delay: Foundation Funding and the Creation of U.S. Climate Change Counter-Movement Organizations. *Climatic Change*. doi:10.1007/s10584-013-1018-7 (December).

Davison, W. 1983. The Third Person Effect in Communiation. *Public Opinion Quarterly* 47(1): 1–15.

Desjardins, L. 2015. What Does Ben Carson Believe? Where the Candidate Stands on 10 Issues. http://www.pbs.org/newshour/updates/ben-carson-believe-candidate-stands-10-issues/

Desjardins, L., and N. Boyd. 2015. What Does Donald Trump Believe? Where the Candidate Stands on 10 Issues. http://www.pbs.org/newshour/updates/donald-trump-believe-candidate-stands-10-issues/

Dewey, J. 1997. *Democracy and Education: An Introduction to the Philosophy of Education*. Free Press. (First published 1916).

Dunwoody, S. 2005, Winter. Weight-of-Evidence Reporting: What Is It? Why Use It? *Nieman Reports* 59(4): 89–91 .http://niemanreports.org/articles/weight-of-evidence-reporting-what-is-it-why-use-it/

Frumhoff, P. C., and N. Oreskes. 2015. Fossil Fuel Groups are Still Bankrolling Climate Denier Lobby Groups. *The Guardian*. http://www.theguardian.com/environment/2015/mar/25/fossil-fuel-firms-are-still-bankrolling-climate-denial-lobby-groups

Gandy, O. 1982. *Beyond Agenda Setting: Information Subsidies and Public Policy.* Ablex Publishers.

Horlick-Jones, T., J. Walls, G. Rowe, N. Pidgeon, W. Poortinga, and T. O'riordan. 2006. On Evaluating the GM Nation: Public Debate About the Commercialisation of Transgenic Crops in Britain. *New Genetics and Society* 25(3): 265–288.

Howe, P., M. Mildenberger, J. Marlon, and A. Leiserowitz. 2015. Geographic Variation in Opinions on Climate Change at State and Local Scales in the USA. *Nature Climate Change* 5: 596–603.

Iyengar, S., and D. Kinder. 1989. *News That Matters: Television and American Opinion.* University of Chicago Press.

McCombs, M., and D. Shaw. 1972. The Agenda-Setting Function of Mass Media. *Public Opinion Quarterly* 36(2): 176–187.

Nelkin, D. 1995. *Selling Science: How the Press Covers Science and Technology.* Rev. ed. W. H. Freeman and Company.

Noelle-Neumann, E. 1993. *The Spiral of Silence: Our Social Skin.* 2nd ed. University of Chicago Press.

Pearson, A.R., and J.P. Schuldt. 2015. Bridging Climate Communication Divides: Beyond the Partisan Gap. *Science Communication* 37(6): 805–812.

Priest, S. 2000. U.S. Public Opinion Divided Over Biotechnology? *Nature Biotechnology* 13: 939–942. (September).

———. 2013. Critical Science Literacy: What Citizens and Journalists Need to Know to Make Sense of Science. *Bulletin of Science, Technology & Society* 33(5–6): 138–145.

Sturgis, P., and N. Allum. 2004. Science in Society: Re-Evaluating the Deficit Model of Public Attitudes. *Public Understanding of Science* 13(1): 55–74.

Subramanian, C. 2013. Rebranding Climate Change as a Public Health Issue. *Time*, August 8. http://healthland.time.com/2013/08/08/rebranding-climate-change-as-a-public-health-issue/

CHAPTER 2

What's the Rush? Reacting to a Slow-Moving Disaster

Why are we not more concerned, on a day-to-day basis, about the impending disaster—or, more accurately, the impending constellation of disasters—that our changing climate is bringing? Why is this situation not nearer to the top of most people's personal agenda of things to worry about? The problem is not limited to explaining "skeptical" or "denier" beliefs or behavior. The same dynamics that affect them also affect the rest of us. Even those who accept that climate change is here may not take action to reduce their carbon dependence or to advocate for new solutions—whether new policy or new technology. It is as though we have all been waiting for someone else to take care of the problem. But no "someone else" is likely to step up and take charge of this; the effort will require all of us to think, plan, and act to avoid the worst of the future climate scenarios and adapt to the inevitable remainder, even if these only seem like apocryphal visions today. This chapter seeks to explore why so many people around the globe, in the United States and elsewhere and including many of those who understand the processes involved, do not seem worried enough to pursue action on climate.

The earth is warming and human activities are a major contributor—almost certainly *the* major contributor—to this trend. Sea levels are rising, glaciers are melting, seasons are changing, storms are worsening, and species are adapting, moving, or dying out. This means that human populations in low-lying coastal areas in particular are threatened even in the short term—and even in the developed world but especially in the less well-protected developing world. Coast dwellers have much to think about, as do farmers and fishermen everywhere. Yet, slightly longer term, all of us (and our children and grandchildren) will experience the consequences, which could ultimately include food shortages if agriculture should become substantially disrupted, as well as additional flooding, fires, and altered weather patterns including extreme storms and in some areas extreme heat, not to mention the very likely complete disappearance of some forms of life and, eventually, the potential for mass migrations of human climate refugees. None of this is fiction, even if it all sounds too much like a Hollywood disaster drama to actually accept. We are already on the path to such a future.

Historically and in evolutionary terms, human beings have been among the most flexible and adaptable of species. Our abilities to observe and to learn (the ultimate foundations of science) and then to transmit that learning through language and other symbolic communication, enabling us to build on our knowledge over generations and to make and use increasingly complex tools, have been the foundations of this success. We have survived and thrived over a truly remarkable range of climates and geographies. For a time, especially during the Cold War era, our ability to make and deploy some of the most complex and dangerous of our tools—nuclear weapons—actively threatened our survival, but so far we seem to have managed to control even that threat. This is evidence that we can come together to control major risks of great complexity that threaten the globe, including the nuclear threat that was of our own making and once seemed uncontrollable. Whether we can do the same for climate remains to be seen, but we must at least try to change course.

Pollutants from industrialization, including the practice of large-scale intensive agriculture, continue to damage our environment, but we have at least made important progress in recognizing—and to an important extent regulating—these threats as well. And although burgeoning human populations around the world often must live in increasingly crowded and stressed environments—some of them desertified through overgrazing by livestock and other impacts of human activity—the human species survives. While we sometimes encounter (partly as a result of the increasing globalization of human society) the threat of disaster from emerging pandemics like Ebola or Zika, modern science-based medicine has increased our lifespan and continues to keep many lethal diseases in check. We would like to imagine that this will go on forever. We seem conveniently complacent, perhaps because of this very success, about the threats of climate change. We may have become too confident, however.

Understanding "Skepticism" and Inaction

It is difficult for most of us to believe that this earth that has nurtured us so far and that we have so effectively mastered may not continue to nurture us in quite the same way that it has done in the past, or that the human traits of tool-making, reasoning, and re-shaping our environment to fit our needs might, at this very moment, be starting to fail us—or worse, are themselves badly eroding some of our own future prospects. The failure to fully grasp what climate change will mean for us or what kind of action will be required to slow it down (it being already impossible to stop it entirely) isn't limited to "skeptics". We are almost all guilty of not taking sufficient action, or of not taking radical enough action, or of taking no action at all, even if we fully accept the science that says climate is changing and we are the problem. We have other problems: unemployment, job loss, threats of terror, shorter-term environmental crises, and for many, coping with an inadequate healthcare system. No major party candidate has run for U.S. president as the "climate candidate".

There are both sociological and psychological reasons why acceptance and change are difficult. For the most part, our tool-making and our ability to reason and communicate seem to have served us very well, and we have continued to advance. Modern science and engineering combine the environmental manipulation possible through the use of extremely sophisticated tools with the advanced use of new symbol systems, such as the languages of mathematics and computer code. In short, we are a very smart species, and this has paid off. However, our system is not a self-correcting one. Functionalism, which was a major sociological approach for over a century, has been largely abandoned by contemporary scholars because it failed to account for *dysfunction*. Society can be compared to an organism with many specialized parts (cells, organs, limbs—or in the case of society, institutions and norms) that work (or function) together. This was believed to maintain social stability, and to a considerable extent that is a true proposition.

However, the central critique of functionalism stresses that society is not really stable at all but is rife with conflicts involving unequal distributions of power, pressures from which are ultimately the source of social change. The social system does not automatically correct itself when it fails (as opposed to the dynamics assumed by the functionalist ideal of stability). Our failure to address climate change sooner rather than later is evidence of dysfunction, and like many social dysfunctions, is related to the distribution of power in society. The fossil fuel industry, the existing governmental structure, and the wasteful energy-dependent lifestyles of a large share of U.S. citizens (and those of many other modern nations) are all strong forces that tend to inhibit change. The solution will not be simply to persuade people that climate change is a problem. We must go further and motivate ourselves to take action that will ultimately result in social change.

Psychologists refer to the tendency to reject information inconsistent with our core beliefs as "cognitive dissonance", an idea generally attributed to Festinger (1957) and that is consistent with the psychology of climate "skeptics" (Lorenzoni et al. 2007). For example, learning that a close and dear friend is actually a liar and a thief would be difficult; we would tend to reject the message, and perhaps we would even blame the messenger. In the same way, learning that life on earth as we know it is threatened is quite frightening, and even if we believe this could be the case, we still tend to discount or ignore the message. Cognitive dissonance—the unpleasant feeling that our cherished core beliefs are being assailed, tempting us to take refuge in rejecting (or at least ignoring) the threat—is a reaction shared across climate believers and "skeptics" alike. For some people, this means they reject the whole idea of climate change. The rest of us may accept the scientific reality of climate change but try not to worry about it, feeling there is little we can do about it—or else we do little or nothing simply because more immediate day-to-day needs take priority. We have families to care for and jobs to do. For now, life goes on. Actors driven by political, economic, and ideological motives exploit our busyness, our complacence with our success as a species, and our unease with considering that this success may already be in decline. They do this by assuring us that climate change doesn't really exist. We are all susceptible to the pull of this seduction, even if we are climate "believers".

Just as some religious fundamentalists may fully understand the idea of evolution but cannot accept it because it conflicts with their deepest beliefs about the universe, we can fully understand (and even accept) the science of climate but stop short of embracing the need for reform. Political actors, meanwhile, are busy trying to provide what they think we need: more energy, cheaper energy, more secure energy rather than cleaner and "greener" energy and reduced consumption. The problem is no longer merely the existence of "skeptical" perspectives. The problem is all of us.

Even if we are not tempted to dismiss the whole idea, we would like to believe that science and technology will conquer this threat, as it has so many others. However, while research continues to explore the possibilities, there is unlikely to be a simple "technological fix" or engineered solution for climate change. We are approaching—and may even have reached—the limits of our technological capacity to resolve the problems created, in large part, by that selfsame technology. And now the second key element of *Homo sapiens*' unique repertoire of abilities—our ability to communicate through language and other symbols—seems to be failing us as well. What science knows to be the certain reality of climate change appears to be escaping the attention of key audiences; even audiences who accept the science are not necessarily prioritizing change. Can improved communication overcome this problem? Can communication research—another kind of (social) science—help to identify what improvements will be effective? To be sure, climate is much more than a communication challenge. But we hope that effective communication about the problem can at least help surmount our collective tendency to avoid it—and it might even motivate real change.

Part of the solution lies in fundamentally reshaping our personal energy usage and national (even global) energy policies. But Americans in particular are used to doing whatever they like, if they can afford it. The necessary steps can be seen as radical and as themselves constituting threats—to our lifestyles and our cherished freedom to do what we please, and also (in the eyes of some) to our economy. And those in developing countries may feel that now it is their turn to enjoy a similarly wasteful lifestyle. Some of those who (for whatever reasons—including both perceived self-interest and their own susceptibility to cognitive dissonance) want to actively delay collective action on climate are skilled users of the tools of persuasion, often far more skilled than most of our scientists, who are rarely trained in communication skills or tactics. We certainly should not expect the scientific community to shoulder the entire burden of making change happen, but they do need to speak out and be heard—as do the rest of the "believer" community.

Other factors also contribute. Our social network—our immediate social environment—is important (for recent evidence, see Yang and Kahlor 2013). If we are surrounded by others who are going about their daily business without visible concern for the issue of climate change and without speaking about it—or even acknowledging it exists—it is psychologically much easier for us to do likewise than to "buck the trend". It is part of fitting in, a natural tendency for us as social creatures. And at the same time, our media system is changing, partly as the result of technological change and partly as the result of the deep and prolonged global economic recession that we are still experiencing, which together are resulting in a fundamental restructuring of the entire media industry. Journalists have historically been our gatekeepers, telling us what issues are important and demanding our attention to them. This system is failing in favor of an Internet free-for-all that is philosophically quite appealing as the foundation of an information democracy, but often deficient as an effective stimulus for focused attention and action.

The reality of climate change is so scary that we all feel much more comfortable ignoring it; it threatens to overpower us with anxiety, dread, concern, even guilt. This makes it easy for us to attend to the "skeptical" voices wanting to put us at ease. Life being what it is for most of us—a daily struggle to accomplish a thousand small tasks—our attention is also distracted. The more immediate demands of meeting basic physical and social needs, making a living, guarding our health, raising our children, paying our bills, and thus maintaining our places in society and our autonomy all take priority, especially in the midst of what seems an intractable economic recession. Abstract and seemingly distant concerns like the long-term future of the climate of the earth are readily set aside. Anyway, or so people may reason, there's very little we can do about this. "Skeptical" arguments that climate is a natural cycle can easily take hold, under these circumstances. And even among the majority who accept the scientific reality of climate change, taking action against it is another matter altogether. We may think that there is nothing we can do, or that someone else will do it.

Change in the Information System

Today, there are fewer journalists working for traditional news organizations and more messages being generated (often via the Internet) by a variety of individual and institutional interests, ranging from activists to corporations and from scientists to ideologues. With an on-the-ground organizational infrastructure of trustworthy advocates for action on climate change largely lacking, conservative corporate interests may dominate; and while many corporations realize they can win public support by acting in a socially responsible way (or seeming to), with public opinion on climate change seemingly divided this may not always be an issue that even the most progressive company wants to take on. Fuel-efficient vehicles seem to sell well, since they appeal both to economic and environmental motivations. But more expensive electricity available through alternative technologies like solar and wind will not be universally popular—and can raise concerns about increasing the cost of doing business and even deepening present conditions of economic recession. Pro-business voices generate information subsidies (press releases, statements, reports, and other instruments created to stimulate a point of view's appearance in news accounts; Gandy 1982) representing these concerns. To be fair, some of these actors may not themselves fully comprehend their own complicity in our climate catastrophe. But complicit they are, even so.

The role of a traditional news organization[1] is to provide real balance among all these competing points of view, to make sure the pro-change voices are heard alongside the anti-change ones. But this kind of balance can be in short supply in our current, economically restricted, media environment. The time and resources available for serious investigative work are limited, increasing reliance on the framings and messages of others. The proliferation of "new media" voices is on the one hand exciting, but the loss of traditional gatekeepers comes at a price. At the same time, superficial or "false" balance in itself has been identified as a problem for climate coverage (Boykoff and Boykoff 2004). Journalists reporting on climate change, many of whom are not science specialists, many very likely unsure at first about the reality of the situation, have routinely fallen back on reporting "skeptical" opinions alongside those speaking for the scientific facts, all in the name of "balance". Both scholars and media critics now refer to this as "false balance" (see Shanahan 2007), and it has receded but it has not disappeared. Some journalists uncomfortable with covering complex science may still be tempted to use this kind of superficial balance as a convenient safety net, however. They are part of the problem as well.

This still goes on, even though (partly as the result of educational efforts aimed at journalists and weathercasters) the "skeptical" voices that used to create "false balance" may be less ubiquitous now. But because the link between extreme weather and climate trends is a probabilistic one, meteorologists, climate scientists, and other experts are often professionally reluctant to attribute specific weather events to the dynamics of climate change. Likewise, journalists and weathercasters reporting on horrendous storms such as Hurricane Katrina, Hurricane Sandy, or Typhoon Haiyan, as well as the wildfires in California and Australia, the melting of glaciers and Arctic ice, unprecedented snowstorms in many areas, and other extreme weather events remain seemingly skittish about connecting any causal dots (for discussion, see Carey 2011). Further, in the Internet world, people can readily find support for any particular point of view, no matter how far out of the mainstream. Whatever your belief, you can find company and comfort there; we can pick and choose our own news sources to reinforce our own preconceptions, leaving us feeling confident rather than unsettled—and less comfortable—in our thinking.

To date, communication seems unable to overcome our complacency or to fully persuade us of the urgency of this slow-moving crisis, even when it persuades us one exists at all. Even if we accept the evidence, some people may reason, isn't this something our grandchildren can deal with, using the science and technology of their future time? Alas, of course, in reality for them it may indeed be too late. If fully accepting the nature of our climate crisis is psychologically uncomfortable, if some of those in our circles of family, friends and co-workers reject it, if we can also find plenty of apparently credible sources on the Internet and elsewhere that reject it, if journalists and trusted leaders are not urging us to rethink our complacency, if we would not know where to turn for advice or how to take action even if we accept that climate change is happening and that we are the root of the problem, and if the change—while not as slow as we might like to think—is not always immediately visible in front of us, then it *makes sense* in a way that so many otherwise rational people are either not very worried about climate change or reject the idea altogether. Scientists in particular, trained to be driven by facts, tend to keep repeating those facts, forgetting that others do not look at the world in quite the same way that they do. The facts do not always help. Of course they are important, but they are just not enough.

COMMUNICATION RESEARCH: CONTRIBUTIONS AND LIMITATIONS

Climate change provides a lens—a critical case study—through which to better understand the relationship between communication and the instigation of action, both individual and collective, in the face of pressing necessity. This relationship has not always received the attention it deserves. Thinking about the reasons we have difficulty confronting the reality of climate change tells us a lot about ourselves as a society and a species. It also tells us something about how communication can (and sometimes cannot) effectively attack other major societal problems. And climate change *is* a social problem, not just a scientific one. That is why it will need a social movement solution.

While hundreds if not thousands of researchers around the world are working to better understand the science of climate change, another cadre of scholars has been working to understand how best to communicate it. As communication researchers ourselves, we argue here that a new paradigm for communication research is essential—one that more fully embraces the inherently *social* nature of human communication.

Just as individual lifestyle adjustments, however essential, are unlikely to be the complete solution to climate change, communication strategies focused on studying and modifying individual behavior, attitudes, and thinking alone—however useful for many types of problems and however valuable as a *partial* strategy for climate change communication—are not by themselves adequate to this broader challenge. And even individual-level change takes place in a broader social context, one in which change can be encouraged or discouraged with important consequences. For example, living in an area where recycling is broadly promoted and accommodated is much different, in terms of individual recycling behavior, from living in an area where it is ignored; within the United States, both contexts can readily be found. It is not clear that recyclers always *know more* about why recycling is important. Rather, it is human nature to conform to shared expectations. Of course, in addition, areas where there is more shared concern about recycling will in all likelihood develop better infrastructure for it—e.g., waste disposal facilities, incentive programs and other policies. The same goes for various forms of alternative energy; not just attitudes but also available technology and other infrastructure often restrict our choices. We live in a socio-technical system, not just a social system.

We do already know a great deal about communication as a uniquely social phenomenon that takes place across cultures and within social networks, at mass as well as interpersonal levels. We know a lot about the relationship between persuasion and social values, as well as about the connection between trust in the messenger and responsiveness to the message. We understand much about the nature of the social institutions and organizations operating in areas such as journalism, advertising and public relations—the institutions and organizations that produce (often in quasi-industrial fashion) messages for our consumption. Communication scholars have studied the relationship between communication and the political system, as well as looking at the political economy of communication institutions themselves. Individual-level studies of the adoption of healthy behaviors, such as quitting smoking, also contribute; they have taught us, for example, that too much fear is not necessarily a motivator, a lesson that has been explored for climate change (O'Neill and Nicholson-Cole 2009).

In urging that we think about things somewhat differently, we certainly do not mean to ignore all of this existing knowledge. We do mean to reconstruct and reorganize it in a somewhat different way for our purposes here, however. Our goal for the present discussion is to consider how all of these forms of knowledge bear on communication prospects and practices for a single case, climate change, rather than singling out only one form or level of communication, as focused scholars tend to do (generally in keeping with the nature of their training). And yet there are also many pieces still missing. Extensive evidence from studies in environmental psychology strongly suggests that moral or ethical norms, alongside attitudes toward environmental behavior, are extremely important (Bamberg and Möser 2007). This is an underappreciated factor in communication studies, both theoretical and applied. Like communication itself, establishing ethical norms is inherently a collective process. And climate change is an issue of social justice.

Distinguishing the Social from the Individual

Some have argued that communication as a research enterprise began in sociology, a field that studies social collectivities and one with which communication studies continues to have many cross-connections. But as Chap. 1 of this book has argued, in contemporary communication research (and especially research on persuasion) the focus has too often been on the individual and too sporadically on the collective. Science communication research (often conceptualized as strategic or persuasive) has absorbed this bias. Rather than disregard our knowledge of the effects of communication on the individual, however, we should seek to better connect it to our knowledge of the social. We should focus additional attention on elements of communication that cannot be so easily studied at the individual level or in the experimental university laboratory and that have therefore been underrepresented in our literature.

This is vital both for enhancing our academic (as well as our practical) understanding of the nature of collective action and for conceptualizing how best to influence it with specific respect to climate change—or any other collective problem. The social cannot always be reduced to the individual; the whole is greater than the sum of the parts. Yet even when communication researchers study *mass* media and their effects, the underlying paradigm often seems implicitly based on a vision of a lone individual sitting (let's say) in front of a television screen, being "exposed" to a message—as though it were a dose of radiation. Experimental investigations that use messages as treatment conditions, often with conveniently available student subjects, generally reproduce this paradigm. The roles of networks of family, friends, co-workers and community in interpreting that message tend to recede into the background.[2] There is no question that experiments, even with students, can produce very useful information. However, this reductionist paradigm cannot readily capture group dynamics, and it is group dynamics that are particularly relevant here. No one can stop climate change by acting alone.

Climate is something that surrounds and influences other activities. Both human civilizations and entire ecosystems survive and thrive within a (physical) climate that nurtures them, or else they fail within a climate that does not. And the particular attitudes and opinions of individual human beings are encouraged to grow or are suppressed and may even die within a particular (social) climate, a climate of public opinion. It matters to us what other individuals and groups are thinking. When we express opinions, we regularly project what others will think of them—and of us—and are likely to adjust our expressions accordingly, consciously or not. This does not always mean that we hide or repress contrary opinions; some of us tend to speak up about some issues even when we believe we are in a contrarian minority (identified as the "hard core" and the "avant garde" in spiral of silence theory; see Griffin 2008). But whatever the outcome, a decision about whether to express an opinion takes account of the surrounding social environment. The complex feedback loops that influence expression may also influence belief.

Yet, methodologically, it is often much easier to study the opinions of individuals than to characterize the climate of opinion in which these are produced. The climate of opinion is not a simple matter of percentages (individuals added up, as in an opinion poll) but also incorporates perceptions of the status, importance, legitimacy, credibility, and power of those individuals, as well as reflects the visibility of those holding particular views, how strongly particular opinions are held within particular social groups, and individuals' self-identifications with some of these groups and their views. Opinion polling serves the needs of politicians seeking election and product advertisers particularly well because they are ultimately concerned with individual choices and behaviors, whether voting behaviors or individual purchasing decisions. We often seek to reform health behaviors and some times environmental behaviors on an individual basis. Yet to effectively produce change in those opinions, decisions and behaviors, we need to also understand, consider, and ultimately influence the collective social environment that produces them. And to understand (or influence) political and regulatory decision making means taking into account the behavior of multiple collectives, from political parties to advocacy groups and from government agencies to corporate lobbyists. These are not individual-level phenomena.

Our all-important impressions of what other people think are formed both in interpersonal conversation and by our exposure to media representations. Media effects researchers do *not* generally conclude that media opinion dictates public opinion. Yet it is by media exposure that we form our impressions of what those outside of our own immediate social circles are thinking, a very powerful if indirect media effect. Media accounts also establish the boundaries between legitimate and illegitimate or "fringe" positions on issues. For climate change, so-called "skeptical" or "denier" voices—especially if represented as possessed of scientific credentials—have often been legitimized by being treated as co-equal and valid, if dissenting, voices within science. This is an important form of misrepresentation, and it likely does have effects, including the distortion of the perceived climate of public opinion, even among climate "believers".

Journalistic "Objectivity" and the Cultivation of Uncertainty

Even though the media's measurable, direct effects on public opinion may be weak, the climate of public opinion is influenced by journalism and other media accounts in many indirect ways, and this in turn may influence public opinion more often and more profoundly than we realize. Because this is not always a short-term process, it may not be easy to reproduce in an experiment or survey—even the most perfectly designed one. Like the cultivation of other perceptions, notably perceptions of the prevalence of violence in the world, a perception of the climate of public opinion is something that evolves and changes over time. Communication scholar George Gerbner, who introduced the idea of cultivation into the world of communication scholarship, was concerned that over-estimation of the prevalence of violence was something of a self-fulfilling prophecy (see Shanahan and Morgan 1999). If everyone believes that the (social) world is a violent place, then our very defensiveness can make it more so. We do not go out at night. We do not always react to violence with surprise or the deep indignation it deserves. Maybe we buy a gun for self-defense, very likely further exacerbating the problem. Gerbner himself worried that these dynamics (what he termed the "mean world view") would cause us to cede more and more power to the police. In many U.S. cities, this prediction has come true: Police are armed with military weapons intended to be used against their own people.

In the United States and many other areas of the world, journalistic "objectivity" often means covering "both" sides of every issue, as though each issue had exactly two sides. Covering both sides may be seen as a key element in creating a "balanced" or fair story. Of course, as discussed above, this is a superficial definition of balance (and many journalists are much smarter than this, to be sure). But it is nevertheless a common and well-respected practice to seek out contrarian opinion to "round out" a story, even in reporting science. This is not a bad practice under many circumstances. But for climate, longer term, this has cultivated the perception that climate "non-believers" represent a legitimate point of view. Even for those who disagree (that is, for the substantial majority of us who do believe in climate change), we may be hesitant to dismiss climate "skepticism" entirely—or to challenge our neighbors' beliefs if they embrace it. And if we ourselves are "skeptical", we are reinforced in the idea that ours is a perfectly valid alternative point of view. Thus the cultivation of uncertainty through the practices of journalism, even science journalism, may also be a self-fulfilling prophecy—just as Gerbner predicted for the cultivation of a perception of the world as a violent and "mean" place.

This phenomenon of covering both sides embodies a tradition that is likely derived from political reporting and is arguably most prominent in nations where there is a clear division between major political interests on the left and on the right. In the United States, for many decades we have had just one major political party on each side; all recent attempts to create a substantive and viable third party (including a "green" party) have so far been unsuccessful.[3] News media structures often reflect political structures. Countries where there are numerous political parties may have more opinionated and diverse news outlets (Italy, for example); countries where there is only one political party have more monolithic news systems (China, for example, although the media situation there has grown substantially more complex in recent decades). In countries with a U.S.-style bipartisan division between two major political parties, such as the United Kingdom, "objective" news is generally thought of as news that includes both corresponding positions. The interpretation of objectivity as left–right balance is well established in these nations.

Given the emphasis on "objectivity" in this sense in U.S. journalism, the incorporation of seemingly well-credentialed climate change deniers in a climate-related story may seem natural and can save the journalist's worrying about accusations of bias. It seems "safe", in other words, and also makes for a nice element of conflict, something that appeals to audiences and therefore to both journalists and news editors. It is also a convenient "way out" for the journalist who is not sure where the scientific consensus actually lies. But in terms of the climate of opinion, this practice undoubtedly has contributed to the demonstrably false perception that there is widespread disagreement within science about climate change.

At the seeming opposite end of the spectrum, another widely recognized but very misleading journalistic pitfall is presenting the results of a single study as representing scientific "truth". For climate change, this can mean that isolated individual studies showing that (say) global warming seems to be slowing down or that non-anthropogenic causes for climate trends predominate can be reported as simple truths. This reflects misunderstanding of the nature of scientific consensus, which is an important aspect of the particular social nature of science itself (see Chap. 6). However, much science news reaches the journalist in the form of a press release from an individual institution whose public relations staff is charged with promoting the research that institution is generating. Single-study science stories are most often the result of single-study press releases, reflecting the institutional nature of news production as universities, research institutes, and even scientific societies seek publicity for their research products—and in the present economically stressed media climate, fewer journalists have the time or resources to investigate these for themselves.

Just as the public opinion climate is influenced by the range of opinions legitimized in media accounts, the climate of opinion within science—as it is seen from outside the scientific community—is also very much a media construction. (The extent to which scientists themselves are influenced by this has been incompletely studied, but it seems likely they are not immune—especially when considering developments outside their own particular specialization.) This is all the more true given that most individuals have no obvious means to verify the truth for themselves, unless they happen to have a scientist friend (who may or may not be a climate specialist).

The social environment matters to opinion formation. News and information also matter, but we interpret that news and information through social processes. When Hurricane Katrina hit New Orleans and surrounding areas in 2005, some demographic groups mobilized more quickly than others. The groups that lagged behind tended to be those in poorer or minority neighborhoods. But, importantly, knowledge per se was not the main issue (Taylor et al. 2009). Those who hesitated did not necessarily lack knowledge of hurricanes or information about the storm, readily available from media sources. However, while they may have been characterized by strong internal social networks, some of these neighborhoods may have lacked equally strong ties to outside networks. This may have limited their awareness of emerging concern elsewhere about the severity of the storm, contributing (along with a lack of access to the resources needed to relocate, concerns about leaving their belongings behind, and other factors) to slower action.

What ultimately triggered action among affected people in New Orleans despite these factors was the influence of neighbors, friends, and family—alongside trusted political leaders speaking through the media—collectively saying, "It's time to go". This mobilization represented a shift in *collective* thinking that (for a variety of reasons) unfolded slightly more quickly in some segments of society than others. In other words, knowing the facts alone was not enough to mobilize affected citizens to action—even in a much more immediate communication emergency of a type that was familiar to many of the affected citizens. New Orleans had experienced hurricanes before; what was initially missing was an understanding of how Katrina was different.

Scientists as Key Communicators; Non-Scientists as Audience

Many scientists, even those who are great teachers, may not be skilled as persuasive communicators; even though trust in scientists in our society is high, that may not be enough, just as the scientific facts on climate change are not enough. Yet scientists' voices clearly need to be heard on the issue of climate. Many scientists teach and are most accustomed to teaching college students, either advanced science students who are already "on board" with the values and much of the vocabulary of science or other students who are required to take science for their degrees and are a captive audience for classes designed for that purpose, which often stress rote learning. Well-prepared, ideally even entertaining, lecturers whose tests cover facts that can be memorized may get high marks as teachers from these students. This doesn't mean that the students are understanding very much about how science actually works, even in their laboratory classes. And when these good lecturers speak to audiences of other academics or even (on relatively rare occasions) to community groups, they are likely preaching to the converted in the sense that these audiences, volunteers rather than conscripts, already find science interesting. None of these audiences really need to be won over to the value of scientific evidence.

It is thus perfectly natural that scientists, including those with the best communication skills for what they are ordinarily called on to do, usually try to address "the problem" of science-related public opinion by more explanation of the science. This doesn't always help, and it is usually not enough. Engaging science teachers are rare and valuable. Those willing to speak to groups outside the university are even more rare. And really interesting media treatments that feature these engaging teachers, while they exist, are not very common—and again are arguably most popular with those who already most interested in science. Popular television science shows featuring science, done well, can help keep science in the public mind, as can science themes in journalism, film, novels, entertainment television, magazines and other popular media—even on YouTube (Allgaier 2013). Climate change discussions can be fostered online and in person, through both government initiatives and protest activities, and via both art and a broad range of alternative media (see Carvalho and Peterson 2012). In other words, the voices of science can be heard in many ways. But even this may not be enough.

Our educational systems do not do enough to foster "critical science literacy", a point this book will return to in Chap. 6; that is, these systems do not reliably provide people with the conceptual and analytical tools they need to form reasoned opinions about issues that rest in part on a foundation of scientific inference, interpretation, and an inevitable lack of total certainty. This is not to say that everyone needs to be a scientist, any more than everyone needs to be a novelist or an artist to think critically about literature or art. But citizens in a science-and-technology-intensive society need more than a collection of facts, many of which quickly grow stale and some of which actually turn out to be wrong in the years after these citizens have left school and become voters. A more complete understanding of "how science works" as a social enterprise would be more enduring, as the years pass. This has to include understanding that the "best available evidence" is all we ever have to go on.

And because everyone is not a scientist, patterns of trust also matter a good deal. (This is true within science as well, given the narrow topical focus of most scientific specialties.) Like other aspects of public opinion, trust is highly social in nature. These patterns vary from culture to culture as well as from individual to individual (Priest et al. 2003). Americans in general tend to trust industry more than members of some other societies, who sometimes exhibit higher levels of trust for environmental voices. Our collective reaction to climate change inevitably reflects these same dynamics.

Ulrich Beck (1992) described modern societies as "risk societies" in which the management of various forms of risk is a central problem—almost an organizing principle for present-day bureaucracies. Despite the complexities, adequately complete and reasonably accurate scientific pictures of some of the risks of modern life have emerged, and these are no longer actively contested: Smoking causes cancer, sendentary lifestyles promote disease, people should wear seat belts while driving, and some chemicals we produce are potentially dangerous and must be regulated or restricted. What will it take to move the risks of climate change from assertions that are seen as quite tentative to things that are accepted as fact? It may take time that we do not really have.

Fortunately, despite the challenges, *most Americans do believe that climate change is occurring.* This is what will allow us, as a society, to move on and concentrate on encouraging collective action toward policy solutions, in cooperation with the rest of the globe. Our apparent political polarization need not indefinitely delay action. Communication (and research about communication) needs to focus on how to make this happen.

Can the Climate of Opinion Actually Be Changed?

While a majority of people in the United States believe in climate change, and even more want energy-related policy change, there seems little capacity for the kind of political action that will make this happen. The current destructive ideological gridlock in the U.S. political system is well known. There are many other challenges to change, including dependence on media for understanding the predominant scientific consensus, cognitive dissonance regarding the implications of a shifting climate (that is, inconsistency of the idea with other, prior beliefs), the demanding nature of modern life, and—among communication researchers—an over-emphasis on understanding the psychology of individuals as opposed to understanding collective social dynamics.

How do we change the climate of public opinion in such a way that collective action on climate is accelerated fast enough to alter in important ways the conditions future generations will find on earth? A huge and diverse society like that of the United States cannot easily be turned in a new direction. Add to that the challenges of acting in concert with the rest of the world, and the problems may indeed seem insurmountable at times. Yet it takes only a glance at the history of successful social movements on such major issues as the implementation of minority civil rights, the establishment of women's rights, or the genesis of the environmental movement to recognize these as instances in which opinion that was once part of the "radical fringe" became increasingly mainstream—that is, increasingly legitimized. None of these battles is yet completely won, to be sure. But change does happen. The issue is whether it can happen fast enough to avoid a global climate catastrophe—or at least to minimize it.

In the history of the environmental movement, for example, initial concerns about pollution, resource waste, and habitat destruction moved from something associated with a handful of extremist organizations to a regular part of more moderate thinking and ultimately to institutionalization in the form of new government regulations and bureaucracies such as, in the United States, the Environmental Protection Agency. This took place over a period of decades rather than centuries. Rachel Carson's environmentalist monograph *Silent Spring* was published in 1962; the Environmental Protection Agency (EPA) was created (by Richard Nixon) in 1970. We would not claim that the EPA is perfect, by any means; like all government agencies, it is influenced by politics and constrained by legislation. Many environmental problems persist and new ones constantly arise, in the United States as elsewhere. Yet we are doing better. For example, the infamous Cuyahoga River in a major industrial district in the state of Ohio, which in 1969 actually caught fire as a result of its severely polluted condition and thus became an iconic symbol of environmental contamination and neglect, is now well on its way to recovery. It is no longer radical to recycle. But unfortunately climate change is progressing rapidly and the natural course of major social change is a slow one.

Climate change is also a much broader problem than past environmental issues, and one of the problems here is that our existing societal infrastructure—our existing constellation of governmental and private institutions and organizations—is organized primarily around other, often more specific, sets of concerns. The EPA has been charged with regulating CO2 emissions, but this was not immediately implemented pending federal court review. Climate change is not just about the environment, not just about energy use, and not solely a problem in international relations. It is all of these and more.

Research-based communication strategies are at least a partial solution. We know a fair amount how to get people to change unhealthy behaviors, even though these campaigns are not always successful in individual cases. To some extent we think we know how to persuade people to vote a certain way, or at least we can often predict these behaviors, and we certainly seem to know how to get people to buy things. We know that communication patterns tend to follow the patterns of social networks and that while mass communication is important, interpersonal communication often remains more persuasive—the combination of the two is especially effective. And we know that people's individual opinions are responsive to the perceived "climate" of opinion in which individuals find themselves embedded, a perception that reflects both mediated and face-to-face communication.

We know somewhat less about how to get people to come together to agree on solutions to pressing problems facing society as a whole, especially problems like climate change that require relatively speedy collective action leading to effective policy solutions. Research on communication for social change has in some ways only just begun. Yet we do already know a good deal, based on communication research to date, about how to persuade people of the basic realities of a changing climate. Key findings from this work are summarized in the next chapter.

Notes

1. This does vary among different cultures, nations, and political systems, although it appears that U.S.-style left-right "balance" has become a more common international norm. That being said, at the same time a number of U.S. cable news networks such as MSNBC on the left and Fox News on the right have grown noticeably more partisan.
2. Of course, it would be an overstatement to say these are *never* studied, but these processes do not always get the attention of scholars in proportion to their manifest importance.
3. The 2016 U.S. presidential election may ultimately be what changes this balance, although it is unlikely to yield a major "green" political party or to elevate climate to a more important issue.

References

Allgaier, J. 2013. On the Shoulders of YouTube: Science in Music Videos. *Science Communication* 35(2): 266–275.

Bamberg, S., and G. Möser. 2007. Twenty Years After Hines, Hungerford, and Tomera: A New Meta-Analysis of Psycho-Social Determinants of Pro-Environmental Behavior. *Journal of Environmental Psychology* 27(1): 14–25.

Beck, U. 1992. *Risk Society: Towards a New Modernity.* Sage.

Boykoff, M.T., and J.M. Boykoff. 2004. Balance as Bias: Global Warming and the US Prestige Press. *Global Environmental Change* 14(2): 125–136.

Carey, J. 2011. Storm Warnings: Extreme Weather is a Product of Climate Change. *Scientific America*, June 28. www.scientificamerican.com/article/extreme-weather-caused-by-climate-change/

Carson, R. 1962. *Silent Spring.* Boston: Houghton Mifflin.

Carvalho, A., and T.R. Peterson. 2012. *Climate Change Politics: Communication and Public Engagement.* Amherst, NY: Cambria Press.

Festinger, L. 1957. *A Theory of Cognitive Dissonance.* Stanford University Press.

Gandy, O.H. 1982. *Beyond Agenda Setting: Information Subsidies and Public Policy.* Norwood, NJ: Ablex.

Griffin, E. 2008. *A First Look at Communication Theory.* 7th ed. McGraw-Hill. Also available online at www.afirstlook.com/docs/spiral.pdf

Lorenzoni, I., S. Nicholson-Cole, and L. Whitmarsh. 2007. Barriers Perceived to Engaging with Climate Change Among the U.K. Public and Their Policy Implications. *Global Environmental Change* 17: 445–459.

O'Neill, S., and S. Nicholson-Cole. 2009. Fear Won't Do It": Promoting Positive Engagement with Climate Change Through Visual and Iconic Representations. *Science Communication* 30(3): 355–379.

Priest, S., H. Bonfadelli, and M. Rusanen. 2003. The "Trust Gap" Hypothesis: Predicting Support for Biotechnology Across National Cultures as a Function of Trust in Actors. *Risk Analysis* 23(4): 751–766.

Shanahan, J., and M. Morgan. 1999. *Television and Its Viewers: Cultivation Theory and Research.* Cambridge University Press.

Shanahan, M. 2007. Talking About a Revolution: Climate Change and the Media. International Institute for Environment and Development, December. dlc.dlib.indiana.edu/dlc/bitstream/handle/10535/6263/Talking%20about%20a%20revolution.pdf?sequence=1&is%20allowed=y

Taylor, K., S. Priest, H. Fussell, S. Banning, and K. Campbell. 2009. Reading Hurricane Katrina: Information Sources and Decision-Making in Response to a Natural Disaster. *Social Epistemology* 23(3–4): 361–280.

Yang, J., and L. Kahlor. 2013. What, Me Worry? The Role of Affect in Information Seeking and Avoidance. *Science Communication* 35(2): 189–212.

CHAPTER 3

Talking Climate: Understanding and Engaging Publics

Portions of the material in this chapter draw heavily from an unpublished literature review developed by Jessica Thompson, currently at Northern Michigan University, using funding providing by the U.S. National Science Foundation's Climate Change Education Partnership program under a grant entitled "Building Place-Based Climate Change Education through the Lens of National Parks and Wildlife Refuges" (award number DBI 1059654).

In 2009, media scholar Matthew Nisbet wrote an essay posing the question of whether the time had finally come for the United States to address climate. In addressing this question, he remarked that public engagement levels seemed low and that it was necessary to reframe climate so that it would appear more relevant to a larger cross-section of Americans (Nisbet 2009). It would seem now that we are still trying to find the right formula for this, some years later. In this chapter, we review some of the evidence—from various disciplines—about why communication efforts have not always reached their intended targets and about how we could do better in this regard. The first rule of effective communication, to know your audience, certainly applies.

In earlier scholarship on "mass" communication, it was common to rely on the assumption of a "mass" public. When there were only three major television networks in the United States and no Internet anywhere, it almost made sense to think this way. Most people (in the developed

© The Author(s) 2016
S. Priest, *Communicating Climate Change*,
Palgrave Studies in Media and Environmental Communication,
DOI 10.1057/978-1-137-58579-0_3

world, at least) seemed to have access to about the same daily diet of information about the world, outside of highly local news and information obtained from interpersonal networks. The media system often appeared monolithic; the existence of international wire services like today's Associated Press often produced relatively homogeneous newspaper agendas as well, with the availability of alternative voices "at a click"—as we know them today—largely unfamiliar. But at present, the proliferation of new media, including the Internet, has made the whole concept of "mass" media somewhat anachronistic, a point we will develop in Chap. 5 in particular. Together with the increasing recognition and valuation of human diversity, this shift has eaten away at the old concept of a "mass" public—terminology that suggests a rather faceless, nameless, and homogeneous group.

For this reason, many of today's communication scholars prefer to speak about "publics" rather than "the public", in recognition of this shift and as a reminder of the importance of identifying key audiences and publics in communication work. Those working from a strictly pragmatic strategic agenda, such as advertisers and political campaign strategists, are well advised to think in terms of important publics or specific key audiences, not "the" public. Targeted marketing is the name of the new game, made easier by the marriage of "big data" and new media technologies. Science-oriented journalists often write (or produce) for a specific set of publics, generally those with higher-than-average interest in scientific topics to begin with. Other science communicators may work for advocacy organizations, universities and colleges, research institutes and think tanks, or science centers and museums, and all of these communicators generally have specific ideas about who they are targeting—and why. Many other factors affect which media and which voices, sources and messages are most likely to reach and influence individual citizens, each of whom is a member of various publics. To a large extent, this complexity in audiences was always there, but it was rather easier to gloss over this diversity when thinking in terms of a single mass public.

Furthermore, many of the relevant publics today for climate change information—as for other issues under public discussion or of high personal relevance—include active information seekers, rather than the sort of passive message consumers that older "mass" media perspectives tended to assume. For example, environmental activists—clearly an important public for climate change—are likely to be active seekers of information about climate and its impacts on ecosystems. Public health professionals may seek information about impacts on the spread of specific diseases. Farmers and gardeners, real estate investors, city planners, those living in coastal areas and others vulnerable to hurricanes, those living in areas subject to wildfires, emergency managers, homeowners and car buyers, naturalists and park rangers, teachers and students—all these and many more seek information for their own specific, often very practical, needs.

Even climate "deniers" may actively seek information about climate science, if only in order to confirm in their own minds that there is good reason to reject it. The "skeptical" public can readily enough find Internet sources that will reinforce their contrarian thinking as well. Research on the nature of risk information seeking and what motivates it is ongoing (for a recent review see Yang et al. 2014). One thing seems certain though: Information, news, and entertainment consumers have ever more choices in front of them, even though we are still learning how they make their decisions among competing sources.

Each of these many publics contributes to the climate of public opinion and how it is perceived. So do public leaders and political groups, and although these are by no means the only influential sources, they may have disproportionate authority. And quite a number of different goals are intertwined with efforts to communicate climate change science to these various audiences, ranging from informal education in hopes of improving our collective capacity for better democratic practice to organizational fundraising for activist groups and from increasing general support for science to providing ongoing news of current scientific developments and commenting on controversial ones—and even providing entertainment (e.g., in fictional films and some documentaries). This creates an eclectic and sometimes confusing mix of messages to approach analytically, and a concomitantly diverse set of research literatures. It also has the potential to bewilder members of various information-seeking publics. Under these circumstances it can be difficult to distill generalizable principles of effective communication, but this chapter seeks to offer a few of them. A rapidly growing body of literature (much more than we could hope to fully discuss here) supports this effort.

One of the most central sets of principles is simply to recognize the diversity of publics that exist, distinguish which of them a particular effort seeks to reach, and then decide what goals and strategies will reach them, which is why this chapter began with a discussion of "publics". All effective strategic communicators seek to reach particular audiences, and even journalists generally want to reach people who will appreciate their stories.

But this approach has its limitations, since we cannot usually control who will receive the information we send out—let alone how various publics will interpret it. Even while we might want to focus on a specific public with our message, it is wise to consider how other publics might react to it. The emails that were the basis of the "Climategate" controversy in which it appeared to members of some publics as though climate scientists were trying to manipulate their data are a case in point; what scientists might have seen as a matter of designing effective communication for use among peers, others might readily interpret as designing propaganda.

While understanding alone may not be enough to spur deeper change, improving understanding of the science related to climate is still an important goal. It just isn't the only goal. Emotional factors, trust in the source from which the message comes, and identity with particular groups (for example, religious or political groups) with their own ideas about climate are all influential as well. To be successful in mitigating the worst effects of climate change, we must encourage a variety of behaviors and actions, not just better understanding of the science: participation in energy conservation; adoption of lifestyle changes that facilitate less frequent travel and simplified commutes; and diffusion of innovations ranging from more energy-conserving construction, appliances, and automobiles to more efficient transportation systems and more sustainable industrial and farming practices that generate less in the way of greenhouse gases. And above all we must persuade both individuals and groups to advocate for broader policy change, encouraging a shift to the use of carbon-conserving alternatives in electric power generation and the adoption of policies at the community, state, federal, and global levels that encourage conservation and reduce greenhouse gas emissions in other ways. Climate change does not have a "one size fits all" solution but needs to be approached from many different directions at once, through communication designed for multiple audiences and purposes, as well as through the development of new energy-related technologies and other sustainable alternatives. Improving knowledge is but one step among many, and it will not by itself guarantee success.

In this chapter, then, we synthesize some of the key strands of recent research on public thinking about climate change and how it relates to climate change communication efforts aimed primarily at the individuals who are the members of all of these overlapping publics. Relevant research suggests that in order to effectively engage audiences in climate change action, communicators must make their audiences aware of the consequences of this very large-scale problem (Stamm et al. 2000), but without relying too strongly on fear-inducing scenarios that can cause people to reject the message (O'Neill and Nicholson-Cole 2009). Given the number of competing demands on people's attention, audiences must also be persuaded to see climate as an important issue and one for which responsibility is shared (Patchen 2010), to feel a sense of personal responsibility to take action (Hart 2010), and to believe that this action will be effective (Feldman and Hart 2016). They must also perceive that others are committed to this cause (Bickerstaff et al. 2008; Etkin and Ho 2007); otherwise, people will feel their individual actions are in vain. And to accomplish all of this, communicators and communication researchers must also seek to understand how addressing climate change relates to people's deeply held political, cultural, and even religious values.

Public Understanding of Climate Science

The recognition that climate change is occurring as a result of human activities is widespread within the global scientific community, but as we all know some segments of the population have been reluctant to accept that this is the case (Borick and Rabe 2010; Gifford 2011; Weber 2010). While most Americans do accept the reality of climate change itself, they remain rather less clear about causation. Doran and Zimmerman (2009) present survey evidence that among climate specialists, 96 % believe that global temperatures have risen and 97 % believe that human activity is a major contributor. Yet even more recently, only 48 % of Americans agreed that global warming is "mainly" caused by human activities (Howe et al. 2015), despite longstanding calls for improving related science education (see, e.g., McCaffrey and Buhr 2008). To be receptive to calls for significant changes in lifestyle and energy generation, a variety of publics need to better understand and accept the relationship between climate and human activity—particularly patterns of energy generation and consumption.

The extent to which people understand the consequences of climate change and are personally engaged in the issue is highly variable from individual to individual. This makes a great deal of sense if considered in the context of interest in science more generally. According to the U.S. National Science Foundation (2014), four out of five Americans say they are interested in "new scientific discoveries". Yet the so-called "attentive public" for science (those who are the most actively interested) has been estimated at only around one fifth of the American population by science literacy scholar Jon Miller (2013). A significant amount of research has been conducted over the past ten years on the consequences of this kind of variability, and these studies confirm that there is not a simple solution to changing peoples' thoughts and beliefs about the issue of climate change (Wolf and Moser 2011). Fortunately, this situation does not require that everyone know everything about the underlying science of climate for us to move forward—only that they understand the basics and accept the major implications. Trust in the scientific enterprise may be as important here as knowledge of the science itself, a point this chapter will return to below.

Americans' understanding increased from 1992 to 2009, with people being less likely to conflate ozone depletion with climate change and more people aware that energy use is a major cause of climate change (Reynolds et al. 2010). However, many Americans remained unaware that atmospheric carbon dioxide concentrations had increased as a result of fossil fuel usage. Because both climate change itself and its interaction with human activities are complex, public confusion is somewhat understandable. Some of this confusion may be attributed to prevalent misconceptions regarding how the climate system works, as well as the statistical uncertainties surrounding specific consequences (Weber and Stern 2011). Of course, active attempts to add to their confusion by climate-denier interests (Oreskes and Conway 2010) certainly have not helped.[1]

Many Americans may lack confidence in the methods and models used by climate scientists, and they may test their personal beliefs about climate change against their own personal observations (Borick and Rabe 2010; Li et al. 2011). This is also understandable and underscores the importance of showing people that the climate is changing in their own backyards. The reliance of many Americans on personal observations of climate change over scientific evidence can be problematic, however, because these observations can be easily misunderstood (Weber 1997). Those who believe the current temperature is warmer than normal have been shown to be more likely to believe in climate change than those who believe the opposite (Joireman et al. 2010). Unfortunately, this mode of thinking might predispose people to become less concerned if the weather turns cold and snowy, which is also a part of climate change as historical weather patterns are increasingly disrupted.

On a related note, as other scholars have pointed out, there has also been popular confusion between the terms "weather" and "climate". Weather is the current temperature and condition at a particular point in time; climate is the average or typical pattern of weather over a longer period of time. Weather is understood as "natural", something that develops on an immense scale and that is not subject to human influence. These attributes may have contributed to the perception that climate change, like the weather, is largely uncontrollable (Bostrom and Lashof 2007). The distinction between the two is in fact philosophically subtle and should not be taken for granted. How is it that while individual weather events are largely unpredictable, we can still be certain that (on average) the pattern of climate is predictable—and changing in predictable ways? Grasping this requires understanding something about the nature of probabilities and how these trends are derived. Not everyone will be comfortable with this type of evidence.

Fortunately, there is some good news to consider, as well as the bad. Despite the fact that there remains a wide range of difference between the most scientifically interested American and the least, as well as between the most scientifically knowledgeable and the least, the bottom line is that we are making important if not always speedy progress on communicating climate change basics. Because of the dynamics of the public opinion climate, however, it is important not only to make additional progress but also to make the progress that has already been achieved as visible as possible. When people understand that basic climate change science is widely accepted both within the scientific community and among a diverse range of other important and familiar publics, all of us are likely to be further reinforced in our commitments both to take personal action and to support collective action. People do not necessarily have to take a statistics course to get "on board" with the basic picture.

Science communicators should also continue to stress that while the specific impacts of climate (in the form of weather events, for example) may not be completely predictable, this does not mean that climate itself is a complete unknown *or that there is no relationship between the two*, as news reports sometimes seem to imply (no doubt in response to scientists' explaining that individual specific events cannot be attributed to climate— true enough on one level, but potentially misleading). Expressions of uncertainty are always essential to reporting science accurately, but there is a difference between uncertainty about very specific predictions (such as might be derived from climate models used in planning for upcoming drought patterns or projections of the direction a hurricane might take when it hits land) and uncertainty about the reality that more change is to be expected—another point that deserves stress.

Ideological Commitments Matter: Politics, Worldviews, and Values

We have recognized for many years that attitudes and opinions about science-related or risk-related issues are not based on available scientific facts alone. Among the best-known early work establishing this was Paul Slovic's classic paper in *Science* (1987) on perceptions of risk, in which he sets forth the idea of what he calls the "psychometric paradigm". Popular perceptions of risk levels are related to a variety of characteristics such as whether exposure to the risk is voluntary, whether the risk is seen as a "dread" risk (one offering the possibility of catastrophic outcomes capable of producing high mortality), whether the risk is well understood, and whether it is controllable. It would certainly appear that worst-case scenarios for the impacts of climate change do not afford people much control over whether they will be exposed to the risks; the situation has clear catastrophic potential, ultimately affecting much of the world's population; and while climate scientists may understand many aspects of climate change, average people may not realize or accept this, nor do they necessarily understand a shifting climate themselves. It may not appear to them that the risks are controllable—indeed, what we believe that we can control is increasingly limited. These factors point to a risk that should be perceived as extreme—and so it is, by many observers.

However, to understand public perception of climate, the bigger point may be that none of these are narrowly "scientific" factors. They do not spring directly from interpreting the scientific evidence about characteristics of the risk itself, in other words. Rather, they reflect values-based judgments and social observations (although these judgments and observations may not always be conscious or explicitly stated). This should not at all be interpreted to mean that people are necessarily "irrational" in their responses to risk, although reactions to climate change also have complex emotional components. Rather, we intuitively take social factors into account in making judgments about the severity of a risk. If more people seem likely to be affected, if the consequences seem likely to be severe, and if we seem to have little collective understanding of or control over a risk, then our concern will be greater. The specific influences in Slovic's original work are subject to revision; a different study asking different questions might yield a different list. But the bottom line is that factors other than the science itself do influence public attitudes.

This makes sense, although it also makes sense that for climate, these familiar indications of risk severity (familiar, that is, to social psychologists and other risk perception researchers) may cause quite a few people to react with denial. Research on this latter phenomenon (how exactly people become "deniers") is not easy to generate, and we do not yet have a complete picture of this.[2] However, other research very clearly supports the view that popular judgments about climate in the United States and elsewhere are influenced less by levels of education or knowledge of the scientific evidence and more by personal beliefs, worldviews, and values (see, for example, Kahan et al. 2011; Poortinga et al. 2011). Climate change has become a politicized and contentious issue in the United States and elsewhere (McCright and Dunlap 2011; Weber 2010; Weber and Stern 2011). And supporters and funders of "denier" perspectives (as well as political candidates and occasional "maverick" scientists, some of whom could well be seeking that kind of support) seem more than willing to continue to exploit this polarization.

The risks posed to human well-being as a result of climate change, much like the negative ecological consequences of increasing global temperatures, are also simply difficult for people to conceptualize. Reser and Swim (2011) see climate change risk perception as a consequence of people's "threat appraisals", the cognitive processes that entail evaluation of potential negative impacts on themselves and on their friends, families, neighbors and co-workers. This approach also emphasizes social factors. These threat appraisals include consideration of possible costs, including psychological ones, as well as the potential physical impacts of a given threat, and they can be affected by communication factors such as conversations with other people and media coverage, both of which contribute—synergistically—to the opinion climate. As risks are reported and discussed, in the media and elsewhere, they can become socially amplified or socially attenuated (Kasperson et al. 1988) as a collective climate of opinion emerges. We do not yet fully understand how one risk becomes popularly overstated or, conversely, ignored (in comparison to the perspectives offered by expert risk assessments), but being able to grasp climate's perceived societal impact "close to home" is clearly an important factor. For climate, both amplification and attenuation seem to be taking place, depending on what publics we are considering.

A number of studies confirm that individuals' partisan affiliations and their ideological perspectives are related to their views on climate change (Hamilton 2011; Malka et al. 2009; McCright and Dunlap 2011; Zia and

Todd 2010). Indeed, at this point this is hardly surprising. More subtle and a bit more surprising: Studies on the politicization of climate change have found that for Democrats, high levels of education and self-reported knowledge of climate change are positively related to belief in and concern about this issue, whereas for Republications high levels of education and self-reported knowledge of climate change are negatively related to belief and concern (Zia and Todd 2010; McCright and Dunlap 2011). The divide between Republicans and Democrats over climate change has become more substantial over the past decade, even though it is also important to remember that many Republicans do accept climate change. In short, the portrayal of climate change as a controversial issue within science involving a high degree of uncertainty seems to have led some people to process information about it in a way that is largely dictated by their political affiliation (Krosnick et al. 2000; Wood and Vedlitz 2007). People often rely on the opinions of trusted political leaders when they perceive that there is a great deal of conflicting information about a given issue, and climate is clearly no exception. So perhaps this outcome is not entirely surprising after all.

Nevertheless, gridlock is not an inevitable outcome, nor is it one we have no way to address. Pearson and Schuldt (2015) recommend that to get beyond this "partisan gap", we pay more attention to group influences *other than* political affiliation, including racial, ethnic, and cultural identities, and modify our communication strategies accordingly. Reshaping the representation of this issue, which has become deeply entangled with various political agendas (Carvalho 2010; Moser and Dilling 2007), will continue to require an ongoing focused effort from climate change communicators to transform the climate change discourse. Demographic factors such as age, gender, and ethnicity are also related to people's knowledge of and concern about climate change (McCright 2010; Wolf and Moser 2011); politics is not the only influence. However, political affiliation is the strongest single demographic predictor of how people engage with this issue (McCright 2009; Borick and Rabe 2010). And more recent evidence continues to indicate that both among those who do not believe climate change is happening and among those that do, levels of certainty can be high (Leiserowitz et al. 2014). This is indicative of the ongoing polarization, a situation that will not be easy to transform as we chart a path forward.

The way that Americans think about our society's economic and regulatory systems—linked in turn to political party identification—can also affect how individuals perceive the issue of climate change and the value of implementing mitigation measures. Because addressing climate change at a large scale will require new regulations, incentive systems, or other means to force reductions in greenhouse gas emissions and thus alter the current system, those opposed to this kind of intervention on ideological grounds (such as anti-big-government conservatives) are less apt to support climate mitigation efforts (see Gifford 2011). Fegina et al. (2010) use the term "system justifiers" to describe those who defend and justify society's status quo. Conservatives and Republicans are more likely to be system justifiers, while liberals, scientists, and environmental groups tend to be more willing to critique the established system (see also McCright 2011).

Yet there are still opportunities here for change. People living in areas of the United States that have already begun to experience changed weather patterns such as decreased rainfall or severe storms are more likely to show concern about climate change than people living in other areas, *in some cases regardless of political affiliation*, according to work by Borick and Rabe (2010). Their study showed that the proportion of Republicans in Mississippi (a state that had recently dealt with extremely powerful hurricanes) who reported that hurricanes increased their belief in climate change was substantially higher than for either Republicans nationally or for Democrats. In interpreting their findings, these researchers argue that Republicans' reliance on personal experiences and observations is consistent with general skepticism about both the media and government. Regardless of the mechanism, though, this study provides intriguing evidence that "denier" beliefs can be effectively challenged by actual experience. This research therefore reinforces the idea that connecting climate to direct experience of observable real-world impact is a vital communication strategy.

Belief systems or ideologies are not just political in nature; they are also, of course, religious. Many studies have demonstrated that people's concern about climate change (as for other societal issues) often reflects their faith-based beliefs (Hayhoe and Farley 2009; Wardekker et al. 2009; Wilkinson 2012). People's religious beliefs may determine whether or not they believe that humans are even capable of altering the earth's weather or climate. Many people commonly refer to weather events as "acts of God" (Bostrom and Lashof 2007). Such fatalistic beliefs can clearly impact people's perspectives on energy use and other relevant choices in that "beliefs in a higher power being responsible for weather extremes" means that "people or governments are not perceived as having any control, influence or responsibility for that which is in God's hands" (Wolf and Moser 2011: 560).

Yet the news on religion is certainly not all bad either. It logically follows that religious leaders can be among the highly trusted sources for climate change information, especially for those groups among whom science may be less trusted. Many religious groups, including groups within American Christianity, embrace the concept of human stewardship over the land—which in turn implies a responsibility for environmental protection. Scientists and science communicators interested in increasing awareness, concern and action about climate change could certainly do more to enlist the help of trusted religious leaders to reinforce this message.

Neither political affiliation nor religious identity fully explains reactions to climate change, and although each of these commonly invoked explanations does have influence, neither one should be viewed as a monolith. There are opportunities as well as challenges involved. Other factors that are particularly important include patterns of trust, a belief that action matters (perceived efficacy), and a sense of responsibility. Each of these offers opportunities for improving communication that could motivate a commitment to change—and thus offers opportunities for new directions in communication research as well that might help move us beyond too exclusive a focus on the present political polarization over climate (as for many other matters).

Trust and Efficacy

This discussion would be incomplete without further considering issues of trust and beliefs about efficacy, two key concepts from risk communication studies. Patterns of trust are very well-established in the risk communication literature as explaining perceptions of a wide range of risk-related public issues. For example, differences in attitudes toward different forms of biotechnology among individual European nations and between Europeans and Americans, which are not well-explained by educational or knowledge differences, are closely related to national patterns of relative trust in key actors such as environmentalists, farmers, or representatives from industry or government (Priest et al. 2003). The patterns that can be observed for *differences* in trust between binary pairs of actor groups (such as environmental groups versus industry) seem to be the key explanation; these differences are generally better predictors of cultural attitudes than the absolute levels of trust in each group. For example, more trust in industry compared to trust in either environmental or consumer organizations seems to explain Americans' greater support for food biotechnology, as opposed to the picture in Europe (Priest et al. 2003: 761).[3]

The picture suggested by this analysis is an intuitively appealing explanation that suggests information consumers are "weighing up" competing arguments based in part on relative trust in different actor groups as they make up their minds about an emerging controversial issue. Another way of thinking about this is that trust may be used as a "heuristic cue" or marker to help sort out whose views can be believed—serving as an "unobtrusive motivator" of behavior change that can operate largely without conscious attention (Dunwoody and Griffin 2015). As we navigate the landscape of available information, in other words, trust seems to be an important signpost at key junctures. As members of an information-rich society, arguably the most information-rich society the world has ever known, we are all subject to information overload and we need such cues to help us navigate the information landscape—likely more now than ever before. Siegrist and Cvetkovich (2000) provide evidence that trust is also more important when people are the least knowledgeable about a hazard.

Individuals' trust in experts and the information they offer has been shown to play a key influential role in shaping people's views about climate change, as it does for a host of other risk-related issues. Malka et al. (2009) confirmed that for people who trust scientific information, increased knowledge about climate change is associated with increased concern, although this is not the case for people who are skeptical of scientists. Further, people who expect scientists to use persuasion are more receptive to that kind of persuasion, although the opposite is true for those who believe scientists' role is simply to inform (Rabinovich et al. 2012). In short, scientists can be effective as trusted communicators, but other messengers may be more effective with some audiences, both those who do not trust science and those who do not expect scientists to act as strategic communicators.

Communication scholars have long recognized that interpersonal communication is at least as important as mass mediated communication, but more research still needs to be done on this intersection in the context of science-related issues (see Southwell and Yzer 2007). Recall from Chap. 2 that when residents of New Orleans made the decision to evacuate as Hurricane Katrina began to engulf them, many times it was interpersonal communication or a combination of interpersonal and mediated communication that was the determining factor (Taylor et al. 2009). The persuasive power of social media likely stems from its resemblance to face-to-face interpersonal communication, even where (as for Twitter, Facebook, and other messaging systems) some messages are actually distributed on a "mass" basis. And trust is very likely one foundation for the persuasiveness of interpersonal communication, another question that may deserve more attention from scholars.

It is not just the message but the messenger, then, that matters. People must trust the sources from which they learn about climate change for this information to be effective. On this basis, Villagran et al. (2010) suggest that doctors, nurses and other medical providers could be trusted sources of climate change information. Their results also reveal that patients who have higher levels of health literacy are actually more likely to engage in climate change mitigation behaviors. And research by Lorenzoni and Pidgeon (2006) suggests that the degree of public trust in government and its institutions affects people's support for mitigation and policy efforts. Ultimately, the public's trust in specific messengers and other actors is centrally important for effective climate change communication.

Television weathercasters (as well as those who report on the weather for other popular media) could be one of the most important messengers on these issues. While only one in five of us may be especially interested in science, just about everyone cares about the weather, making the TV weathercaster the only type of science journalist many of us actively listen to (Wilson 2008). Weathercasters are a vitally important group who should continue to find ways to incorporate climate information into their messages; other communicators and communication researchers should continue to keep in mind the key role this group occupies in terms of broad audience reach. Weathercaster educational activity seems to be increasing, and it could provide new opportunities for university-based scientists and weathercasters to work together.[4]

Without this kind of information on present and local effects, it is easier for people to assume that the risks of climate change may all be distant or future (what Gifford has referred to as "judgmental discounting"; 2011). This type of thought process can explain some of the lack of concern and action among many Americans in regard to climate change. One study from 2005 showed that while 68 % of Americans are most concerned about the impacts on people around the world and nonhuman nature, only 13 % are most concerned about themselves and their family (Leiserowitz 2005).

Another, more recent, study yielded consistent findings, demonstrating that most Americans believe people in developing countries are more vulnerable to health risks associated with climate change than are people in the United States (Akerlof et al. 2010). The belief that the impacts of climate change are distant and relatively unthreatening at home is problematic, of course, because this perception can reduce people's sense of urgency about it.

Because climate change is a global issue with far-reaching consequences, people may also tend to feel that they are not capable of affecting it as individuals. They can feel as though others and not themselves are the ones responsible. People feel responsible to act when they become aware of negative consequences that would affect other people if they fail to act; personal responsibility is experienced as a moral obligation to act, and this obligation arises from the activation of an individual's personal moral norms (Weber and Stern 2011). For climate change, though, it can be difficult for people to understand the connection between their own actions and the well-being of others because of the complexity and scale of climate change and its global consequences, which can sometimes feel overwhelming.

Researchers in the United Kingdom have found that people do feel that they are personally responsible for climate change, as well as responsible to take mitigating action (Lorenzoni et al. 2007). Other evidence confirms that efficacy beliefs, which are widely recognized as being important for the adoption of healthy behaviors, are also important influences on adopting pro-social environmental behaviors related to climate change, with a higher sense of self-efficacy associated with a greater understanding of the urgency for action at the collective level (Koletsou and Mancy 2011). According to Feldman and Hart (2016), a sense of efficacy that increases hope increases commitment to climate-related political action. Increasing the extent that people feel a sense of community and believe that their actions are part of a collective effort to mitigate climate change could enhance the extent that they believe their individual actions make a difference (McKenzie-Mohr 2011).[5]

It is for that reason—the apparent connection between the idea of collective action and a sense of personal, as well as collective, efficacy—that this book is interested in understanding in more depth the role of collective discussions, actions and solutions, rather than solely individual ones, and in creating the kind of opinion climate in which climate change action is seen as a shared priority. In this regard the December 2015 Paris

COP21 meeting in which national representatives from around the globe pledged action was a vital step forward, not only because of the universal commitment to national change but also because this commitment was a *visible* one. Nevertheless, the two levels must work together. Collective solutions are not possible without individual commitment, any more than individual commitment can be completely successful without the adoption of collective solutions.

What Have We Learned?

A bewildering constellation of factors matter to attitudes, opinions and beliefs regarding climate: knowledge and information, to some extent, but also ideology, religion, political affiliation, emotions, trust, and efficacy beliefs. These are the lenses through which information is filtered and interpreted. However, this does not mean that people who fall into particular religious or political groups should be seen as the enemies of climate change responses. The truth is much more nuanced. Some highly religious people are committed to responsible stewardship of the Earth; some on the political right are as concerned about environment and even climate as many of the rest of us. Communicating with these and other groups should not assume they are negative about climate but instead might look for places to connect climate concerns as directly as possible with elements of their worldviews—a good idea no matter what public is being addressed.

In the next chapter, we transition from a discussion of persuading individuals directly to a closer look at the changing social conditions that also influence the spread of climate information and awareness. To work toward broad social change, we need to consider both individual and collective processes.

Notes

1. It is difficult to judge just how much they may have hurt. Certainly people uncomfortable with ordinary scientific uncertainty and those uncomfortable with the profound implications of climate change are among those especially vulnerable to these influences (see Chap. 2). However, in order to find a path forward, we believe it is now important to concentrate on promoting positive steps that those who recognize the reality of climate change must take, both at the individual and collective levels, rather than focusing too exclusively on correcting erroneous "denier" perspectives.

2. This point is not easily researched, since it is a practical impossibility to measure how severe specific people would have imagined the consequences of climate change to be *were they not* in denial on this point to begin with. What we usually *can* study is more likely to consist of demographic indicators than observations of the actual emergence of denial over time in more naturalistic terms.
3. This means that for attitudes toward agricultural biotechnology, a high national level of trust in (say) environmental groups is less important alone than if it is combined with a low level of national trust in (say) industry, or vice versa. The ways that trust is influential can be more complex than we might expect, in other words.
4. For example, an interesting March 9, 2015, article in the *Washington Post* (authored by two well-credentialed atmospheric scientists but published under the auspices of the *Post's* "Capital Weather Gang") addressed the complex relationship between warmer temperatures and higher snowfalls in the context of climate trends.
5. Oddly, though, at least one widely cited U.S. study has shown that people who feel more informed about climate change, along with those who have high confidence in scientists' understanding, actually tend to feel less responsible and experience less concern (Kellstedt et al. 2008). Perhaps a sense of being well informed gives some people more confidence that the issue can be (or is being) addressed, or perhaps there is another explanation. More research is still needed on these issues.

References

Akerlof, K., R. DeBono, P. Berry, A. Leiserowitz, C. Roser-Renouf, K. Clarke, A. Rogaeva, M.C. Nisbet, M.R. Weathers, and E.W. Maibach. 2010. Public Perceptions of Climate Change as a Human Health Risk: Surveys of the United States, Canada and Malta. *International Journal of Environmental Research and Public Health* 7(6): 2559–2606.

Bickerstaff, K., P. Simmons, and N. Pidgeon. 2008. Constructing Responsibilities for Risk: Negotiating Citizen-State Relationships. *Environment and Planning A* 40(6): 1312–1330.

Borick, C.P., and B.G. Rabe. 2010. A Reason to Believe: Examining the Factors that Determine Individual Views on Global Warming. *Social Science Quarterly* 91(3): 777–800.

Bostrom, A., and D. Lashof. 2007. Weather or Climate? In *Creating a Climate for Change: Communicating Climate Change and Facilitating Social Change*, eds. S. C. Moser and L. Dilling, 31–43. Cambridge University Press.

Carvalho, A. 2010. Media(ted) Social Discourses and Climate Change: A Focus on Political Subjectivity and (Dis)engagement. *Wiley Interdisciplinary Reviews–Climate Change* 1(2): 172–179.

Doran, P. T., and M. K. Zimmerman. 2009. Examining the Scientific Consensus on Climate Change. *Eos, Transactions American Geophysical Union* 90(3): 22–23.

Dunwoody, S., and R. Griffin. 2015. Risk Information Seeking and Processing Model. In *The SAGE Handbook of Risk Communication*, eds. H. Cho, T. Reimer, and K. McComas, 102–116. Sage.

Etkin, D., and E. Ho. 2007. Climate Change: Perceptions and Discourse of Risk. *Journal of Risk Research* 10(5): 623–641.

Fegina, I., J.T. Jost, and R.E. Goldsmith. 2010. System Justification, the Denial of Global Warming, and the Possibility of "System-Sanctioned Change". *Personality and Social Psychology Bulletin* 36(3): 326–338.

Feldman, L., and P.S. Hart. 2016. Using Political Efficacy Messages to Increase Climate Activism: The Mediating Role of Emotions. *Science Communication* 38(1): 99–127.

Gifford, R. 2011. The Dragons of Inaction: Psychological Barriers that Limit Climate Change Mitigation and Adaptation. *American Psychologist* 66(4): 290–302.

Hamilton, L.C. 2011. Education, Politics, and Opinions About Climate Change; Evidence for Interactions. *Climatic Change* 104(2): 231–242.

Hart, P. S. 2010. Prosocial Messages and Perceptual Screens: Framing Global Climate Change. PhD diss., Cornell University.

Hayhoe, K., and A. Farley. 2009. *A Climate for Change: Global Warming Facts for Faith-Based Decisions*. FaithWords.

Howe, P., M. Mildenberger, J. Marlon, and A. Leiserowitz. 2015. Geographic Variation in Opinions on Climate Change at State and Local Scales in the USA. *Nature Climate Change* 5: 596–603.

Joireman, J., H.B. Truelove, and B. Duell. 2010. Effect of Outdoor Temperature, Heat Primes, and Anchoring on Belief in Global Warming. *Journal of Environmental Psychology* 30: 358–367.

Kahan, D.M., H. Jenkins-Smith, and D. Braman. 2011. Cultural Cognition of Scientific Consensus. *Journal of Risk Research* 14(2): 147–174.

Kasperson, R.E., O. Renn, P. Slovic, H.S. Brown, J. Emel, R. Goble, J.X. Kasperson, and S. Ratick. 1988. The Social Amplification of Risk: A Conceptual Framework. *Risk Analysis* 8(2): 177–187.

Kellstedt, P.M., S. Zahran, and A. Vedlitz. 2008. Personal Efficacy, the Information Environment, and Attitudes Toward Global Warming and Climate Change in the United States. *Risk Analysis* 28(1): 113–126.

Koletsou, A., and R. Mancy. 2011. Which Efficacy Constructs for Large-Scale Social Dilemma Problems? Individual and Collective Forms of Efficacy and

Outcome Expectancies in the Context of Climate Change Mitigation. *Risk Management* 13: 184–208.

Krosnick, J.A., A.L. Holbrook, and P.S. Visser. 2000. The Impact of the Fall 1997 Debate About Global Warming on American Public Opinion. *Public Understanding of Science* 9: 239–260.

Leiserowitz, A.A. 2005. American Risk Perceptions: Is Climate Change Dangerous? *Risk Analysis* 25(6): 1433–1442.

Leiserowitz, A., E. Maibach, C. Roser-Renouf, G. Feinberg, S. Rosenberg, and J. Marlon. 2014. *Climate Change in the American Mind: October 2014.* Yale Project on Climate Change Communication. http://environment.yale.edu/climate-communication-OFF/files/Climate-Change-American-Mind-October-2014.pdf

Li, Y., E.J. Johnson, and L. Zaval. 2011. Local Warming: Daily Temperature Change Helps Influence Belief in Global Warming. *Psychological Science* 22(4): 454–459.

Lorenzoni, I., and N. Pidgeon. 2006. Public Views on Climate Change: European and USA Perspectives. *Climatic Change* 77(1–2): 73–95.

Lorenzoni, I., S. Nicholson-Cole, and L. Whitmarsh. 2007. Barriers Perceived to Engaging with Climate Change Among the UK Public and Their Policy Implications. *Global Environmental Change* 17(3–4): 445–459.

Malka, A., J.A. Krosnick, and G. Langer. 2009. The Association of Knowledge with Concern About Global Warming: Trusted Information Sources Shape Public Thinking. *Risk Analysis* 29(5): 633–647.

McCaffrey, M.S., and S.M. Buhr. 2008. Clarifying Climate Confusion: Addressing Systemic Holes, Cognitive Gaps and Misconceptions Through Climate Literacy. *Physical Geography* 29(6): 512–528.

McCright, A.M. 2009. The Social Bases of Climate Change Knowledge, Concern, and Policy Support in the U.S. Public. *Hofstra Law Review* 37(4): 1017–1047.

McCright, A. M. 2010. Dealing with Climate Change Contrarians. In *Creating a Climate for Change: Communicating Climate Change and Facilitating Social Change*, eds. S. C. Moser and L. Dilling, 200–212. Cambridge University Press.

McCright, A.M. 2011. Political Orientation Moderates Americans' Beliefs and Concerns About Climate Change. *Climatic Change* 104(2): 243–253.

McCright, A.M., and R.E. Dunlap. 2011. The Politicization of Climate Change and Polarization in the American Public's Views of Global Warming, 2001–2010. *Sociological Quarterly* 52(2): 155–194.

McKenzie-Mohr, D. 2011. *Fostering Sustainable Behavior: An Introduction to Community-Based Social Marketing.* 3rd ed. New Society Publishers.

Miller, J. 2013. *The American People and Science Policy: The Role of Public Attitudes in the Policy Process.* Elsevier. (Original published 1983, Pergamon Press).

Moser, S. C., and L. Dilling. 2007. Toward the Social Tipping Point: Creating a Climate for Change. In *Creating a Climate for Change: Communicating*

Climate Change and Facilitating Social Change, eds. S. C. Moser and L. Dilling, 491–516. Cambridge University Press.

Nisbet, M. 2009. Communicating Climate Change: Why Frames Matter for Public Engagement. *Environment: Science and Policy for Sustainable Development*, March–April. http://www.environmentmagazine.org/Archives/Back%20Issues/March-April%202009/Nisbet-full.html

O'Neill, S., and S. Nicholson-Cole. 2009. Fear Won't Do It: Promoting Positive Engagement with Climate Change Through Visual and Iconic Representations. *Science Communication* 30(3): 355–379.

Oreskes, N., and E. Conway. 2010. *Merchants of Doubt: How a Handful of Scienitsts Obscured the Truth on Issues of Tobacco Smoke to Global Warming*. Bloomsbury Press.

Patchen, M. 2010. What Shapes Public Reactions to Climate Change? Overview of Research and Policy Implications. *Analyses of Social Issues and Public Policy* 10(1): 47–68.

Pearson, A.R., and J.P. Schuldt. 2015. Bridging Climate Communication Divides: Beyond the Partisan Gap. *Science Communication* 37(6): 805–812.

Poortinga, W., A. Spence, L. Whitmarsh, S. Capstick, and N. Pidgean. 2011. Uncertain Climate: An Investigation into Public Skepticism About Anthropogenic Climate Change. *Global Environmental Change* 21(3): 1015–1024.

Priest, S., H. Bonfadelli, and M. Rusanen. 2003. The "Trust Gap" Hypothesis: Predicting Support for Biotechnology Across National Cultures as a Function of Trust in Actors. *Risk Analysis* 23(4): 751–766.

Rabinovich, A., T.A. Morton, and M.E. Birney. 2012. Communicating Climate Science: The Role of Perceived Communicator's Motives. *Journal of Environmental Psychology* 32(1): 11–18.

Reser, J.P., and J.K. Swim. 2011. Adapting to and Coping with the Threat and Impacts of Climate Change. *American Psychologist* 66: 287–289.

Reynolds, T.W., A. Bostrom, D. Read, and M.G. Morgan. 2010. Now What Do People Know About Global Climate Change? Survey Studies of Educated Laypeople. *Risk Analysis* 30(10): 1520–1538.

Siegrist, M., and E. Cvetkovich. 2000. The Role of Social Trust and Knowledge. *Risk Analysis* 20(5): 713–720.

Slovic, P. 1987. Perception of Risk. *Science* 236: 280–236.

Southwell, B.G., and M.C. Yzer. 2007. The Roles of Interpersonal Communication in Mass Media Campaigns. *Communication Yearbook* 31: 419–462.

Stamm, K.R., F. Clark, and P.R. Eblacas. 2000. Mass Communication and Public Understanding of Environmental Problems: The Case of Global Warming. *Public Understanding of Science* 9(3): 219–237.

Taylor, K., S. Priest, H. Fussell, S. Banning, and K. Campbell. 2009. Reading Hurricane Katrina: Information Sources and Decision-Making in Response to a Natural Disaster. *Social Epistemology* 23(3–4): 361–280.

U.S. National Science Foundation. 2014. Science and Technology: Public Attitudes and Understanding. Chapter 7 of *Science & Engineering Indicators 2014*. http://www.nsf.gov/statistics/seind14/index.cfm/chapter-7/c7h.htm

Villagran, M.M., M. Weathers, B. Keefe, and L. Sparks. 2010. Medical Providers as Global Warming and Climate Change Health Educators: A Health Literacy Approach. *Communication Education* 59(3): 312–327.

Wardekker, J.A., A.C. Petersena, and J.P. van der Sluijs. 2009. Ethics and Public Perception of Climate Change: Exploring the Christian Voices in the U.S. Debate. *Global Environmental Change* 19: 512–521.

Weber, E. U. 1997. Perception and Expectation of Climate Change: Precondition for Economic and Technological Adaptation. In *Psychological and Ethical Perspectives to Environmental and Ethical Issues in Management*, eds. M. Bazerman, D. Messick, A. Tenbrunsel, and K. Wade-Benzoni, 314–341. Jossey-Bass.

Weber, E.U. 2010. What Shapes Perceptions of Climate Change? *Wiley Interdisciplinary Reviews–Climate Change* 1(3): 332–342.

Weber, E.U., and P.C. Stern. 2011. Public Understanding of Climate Change in the United States. *American Psychologist* 66(4): 315–328.

Wilkinson, K. K. 2012. *Between God and Green: How Evangelicals are Cultivating a Middle Ground on Climate Change*. Oxford University Press.

Wilson, K. 2008. Television Weathercasters as Science Communicators. *Public Understanding of Science* 17: 73–87.

Wolf, J., and S.C. Moser. 2011. Individual Understandings, Perceptions, and Engagement with Climate Change: Insights from In-Depth studies Across the World. *Wiley Interdisciplinary Reviews–Climate Change* 2(4): 547–569.

Wood, B.D., and A. Vedlitz. 2007. Issue Definition, Information Processing, and the Politics of Global Warming. *American Journal of Political Science* 51(3): 552–568.

Yang, Z.J., A.M. Aloe, and T.H. Feeley. 2014. Risk Information Seeking and Processing Model: A Meta-Analysis. *Journal of Communication* 64(1): 20–41.

Zia, A., and A.M. Todd. 2010. Evaluating the Effects of Ideology on Public Understanding of Climate Change Science: How to Improve Communication Across Ideological Divides. *Public Understanding of Science* 19(6): 743–761.

CHAPTER 4

The Evolving Social Ecology of Science Communication

The previous chapter of this book has emphasized the need to understand the thoughts, opinions, attitudes and behavioral dynamics that flourish at the individual level—but, importantly, to understand these in the context of a collective climate of public opinion. The phrase "climate of public opinion" well captures the complex interplay between the individual and the collective: Public opinion responds to multiple influences and may seem to shift unpredictably, but it often reveals meaningful trends over time. While the origins of this phrase may be older, it is often associated—in both communication studies and political science—with the work of Elisabeth Noelle-Neumann (1993), who as mentioned in Chap. 1 argued that we are constantly assessing the climate of opinion in deciding whether it is socially risky for us to voice our own views. In a climate of social risk, a "spiral of silence" can be generated in which more and more people fail to speak up. The media can tend to reinforce such spirals by representing—explicitly through something like poll data or less directly through example, accurately or not—what public opinion is at a given moment. The purpose of the present chapter is to begin to further unpack the relevant processes that operate at a more distinctly collective level.

The United States has a culture with a reputation for being highly focused on the individual. We can see this in both news and entertainment media: Stories of celebrity actors, entrepreneurs, sports stars, and political candidates seem to captivate U.S. audiences; so do stories of highly original (or merely

© The Author(s) 2016
S. Priest, *Communicating Climate Change*,
Palgrave Studies in Media and Environmental Communication,
DOI 10.1057/978-1-137-58579-0_4

highly deviant) criminals. The mundane, day-to-day, workings of the successful production companies, thriving corporations, winning sports teams, and active political parties who actually generate these high-profile individuals seem less entertaining. Stories about the social reasons for crime are nearly invisible; if there is a puzzle about why someone went wrong, it is seen as a puzzle about the psychology of the individual. In short, it is very often the individual that is so often the central focus of attention. As a nation, the United States is not necessarily unique in this regard, but if it were possible for all cultures to be put on a continuum in terms of individualism, we would very likely be close to the "extremely individualistic" end of the spectrum.

Strategic communication research most commonly seeks to influence the individual as well. After all, we are a consumer society, and it is individuals who buy products (presumably in response to advertising messages) and, as citizens, vote for political candidates (presumably as a result of the messages used in their political campaigns). Understanding how individual members of "the public" think about climate, and communicating messages built on that understanding will remain important. However, it is not individuals—acting alone—who will solve the climate change dilemma. It is individuals *acting together*, who know they are acting with many others, who will solve it.

Individual motivations, beliefs, values, emotions, and knowledge are all important, but it is easy to underestimate the power of collective processes when focusing on individual-level factors. We are all heavily influenced by perceptions of what others think—even in the case of people's "uptake" of factual scientific knowledge. Which sources for science are seen as credible, and what science is seen as correct? Such judgments are themselves influenced by what others think, reflecting collective factors such as institutional reputations. Many interdependent groups and organizations—not simply journalists or scientists acting on their own but media organizations, government agencies, nonprofit groups, corporate entities, public relations firms, scientific organizations such as universities, research institutes and academic journals, plus a host of professional societies—make up the social ecology for communicating science, including climate science. Yet journalists and scientists remain key actors.

ORGANIZATIONS AND INSTITUTIONS AS AGENDA-BUILDERS

Organizations and institutions operating in the middle level between the individual and the societal have substantial impacts on both professional activity and individuals' opinions. These organizations and institutions, while of course made up of individuals, can also be conceptualized as actors in their own right. They form a complex and enduring web. The "lives" of organizations generally extend beyond the careers and even the physical lives of their individual members. It is organizations that collectively adopt longer-term policies and purposes and take actions to advance these.

While we may think first of the news media in the context of science communication, there are many other organizations and institutions that are involved with the dissemination of scientific information. Some of these organizations are a formal part of government; others are private nonprofit groups (often referred to as NGOs, meaning non-governmental organizations). Large segments of the corporate world are also concerned with science and technology. And schools, science centers and museums, universities, research institutes, and libraries are also among the key institutions for communicating science, rather than just storing it away.

The characteristics and great complexity of this layer that exists "in between" the individual and the societal is worth thinking about, as many of these organizations and institutions have a good deal of influence on how science, including climate science, is understood and communicated. For example, influence is wielded not just by media employers or by the various professional groups with which science communicators may affiliate, but also by authoritative bodies from government. Considering just the federal layer of government in the United States, a proverbial alphabet soup of well-known agencies either study, monitor, or are tasked with managing climate impacts in the course of overseeing our public resources and our agricultural and energy production. A few of the most important include the DOE, EPA, FWS, NASA, NOAA, NPS, NWS, and USDA,[1] although climate affects the work of all governmental agencies just as it affects all of us as individuals.

Yet government employees, just like the rest of us, may be reluctant to jump into the fray where they feel their personal views might actually get them into trouble, a spiral of silence phenomenon. For climate, because of its politicization, breaking this silence has sometimes come at a political price. Different federal administrations have managed the relationship between federal science and society at large with different degrees of openness. In a widely recognized case, high-ranking NASA climate scientist James Hansen reported that the NASA public relations office (then under the administration of President George W. Bush) tried to silence him on the topic of climate change (Revkin 2006). Hansen was also reportedly the subject of rebuke from "skeptical" scientists, as well as the target of alleged backlash from government sources. The potential for such reactions creates a "chilling effect" for other public servants who might speak out on controversial issues.

And yet all of these government agencies (and a myriad of others, including state and local agencies with somewhat parallel missions) interpret scientific evidence and apply it in their work. Directly or indirectly, this can influence what is considered credible and legitimate science and what is not, whether the agencies actually create that research, interpret it for others, or simply decide how to use it and how (or whether) to communicate those decisions. Planning, transportation, and emergency management agencies at every level need to concern themselves with sea level rise and storm response, especially in coastal areas. Public health agencies are concerned as well because of the potential for changing patterns of disease in response to climate-related ecosystem change and vector species redistribution. Many of these governmental institutions communicate in a very pro-active way (e.g., by creating and distributing press releases and by maintaining informational websites). They help create part of the opinion climate for climate change, but they are not generally in the business of adopting what might be perceived as controversial positions.

Many other organizations and institutions form parts of the social ecology of science and contribute to the climate of opinion for issues like climate change as well. Arguably among the most influential groups in the United States for the evaluation of science, the National Academy of Sciences (NAS), is actually not a government agency at all but a nongovernmental organization operating with federal funding. NAS issues dozens of reports every year on new, emerging, and controversial science-related issues. Major peer-reviewed scientific journals such as *Science* or *Nature* also play key roles in determining what science is considered acceptable and what is not, as (for climate) do climate-specialist journals such as *Climatic Change* or the *Journal of Climate*. Indeed, every scientific journal plays some role, but the top-tier science journals are of course the most influential. Journals sometimes engage in their own public relations and public education efforts, as do major scientific societies such as the American Association for the Advancement of Science (AAAS, the publisher of *Science*).

Organizations outside the United States influence the nation internally as well. A list of 197 scientific organizations worldwide that accept that climate change has been caused by humans has been compiled by the state government of California.[2] Why did they do this? Presumably for the purpose of demonstrating that the consensus on climate change is a global one. Operating at this global scale, the specially created Intergovernmental Panel on Climate Change (IPCC), an organization with broad international membership (195 countries), was formed in 1988 and operates under the auspices of the United Nations Environment Program and the World Meteorological Organization. The IPCC has become an especially important source for climate change information and received the Nobel Peace Prize (shared with Al Gore) for this work; they continually review the latest research on climate and issue painstakingly detailed reports that generally receive a lot of press—and what often seems like unending scrutiny, as well they should. Their operating principles specify they should do this work in a "comprehensive, objective, open, and transparent" way. Certainly the IPCC may not be perfect, but it has undoubtedly helped put climate issues on the public agenda.

If that is not enough complexity, the many hundreds or even thousands of private nonprofit organizations that advocate for environment, wildlife, or public health in the United States and elsewhere can all potentially weigh in on the nature and meaning of scientific evidence—including the scientific evidence surrounding climate change—and so can religious and political groups from all parts of the human spectrum, whether or not they have adopted any official position. And, of course, so-called "climate skeptic" (or "climate denier") organizations can and do weigh in as well. All of these institutions and organizations and many more make up the environment in which information about climate science and climate change is produced, evaluated and communicated. With so many potential sources of information, it is little wonder we are not all on the same page at times.[3]

Media researchers have used the term agenda-setting for many years, since McCombs and Shaw's early work on this issue (1972) first established that what the news media represent as important often seems to translate into what the news-consuming public thinks is important. However, news media do not operate in a vacuum, and they do not set the agenda for public discussion unilaterally. The news agenda is built from the actions of multiple institutions in a broader process sometimes called agenda-building (see Cobb and Elder 1983), a type of collective activity operating primarily at the institutional level. In the case of science, this process might fairly be described as quite complex. Most non-scientist members of the public have few reliable sources they can consult directly on an interpersonal basis if things seem uncertain. Thus, despite the complexity involved in creating the media agenda for scientific topics, what the media ultimately say about that science may be especially influential. Their news agenda may not match the news agenda of major scientific groups, and it does not always match the public agenda either, but all three influence one another. Media reports, whether in print, online or broadcast, remain the key source for most non-scientists of scientific information.

PROFESSIONAL ASSOCIATIONS FOR JOURNALISTS AND SCIENTISTS

Unlike our cultural identities, such as identification with a particular ethnic group or (say) the religion we grew up with, our personal affiliations with most organizational and institutional groups are essentially voluntary. This includes membership in such organizations as political parties, social and public service organizations, and professional groups. For the field of science communication, in addition to the many other organizations that employ science communicators (from media to museums and from universities to government agencies), the professional societies in journalism (and to some extent those in science itself) are arguably among the most influential groups, especially with respect to defining the ethical responsibilities of professionals.

In the United States, as in many other (but not all) countries, the idea of general "certification" or any *required* organizational affiliation for journalists has not been a very popular idea; for some observers, this can be seen as antithetical to the principle of free speech.[4] Anyone, in theory, can be a journalist, although most full-time journalists still work for an organization with control over the work; their freedom to write about what they like and say what they please therefore has limits. However, while affiliation is completely voluntary, professional societies such as the Society of Professional Journalists, the National Organization of Science Writers, and the Society of Environmental Journalists, all operating independently of any individual news or media organization, are quite influential with respect to setting standards for responsible and ethical professional behavior, including journalistic independence.

Scientists as well as journalists embrace professional free speech, which in the case of scientists is often guaranteed—in the form of academic freedom—to tenured faculty, although this is also a kind of freedom within limits. And individual scientists are very often affiliated with one or more scientific societies that represent their specialization; these are often the publishers of related academic journals and the organizers of major scientific conferences, just as occurs in the world of academic communication studies. Interestingly, both scientists and journalists share another characteristic: The individual journalist and the individual scientist often owe allegiance (in many cases their primary professional allegiance) to their profession as a whole, rather than their current specific employer. Thus it makes sense that professional associations in both journalism and science may have quite a bit of influence.

A small proportion of science communicators do their work as volunteers. This would include, for example, a volunteer docent at a science museum, a scientist who participates in science café discussions in the evening, or a community organizer who gets involved in a project about environment in his or her spare time. Roles in such "extra-curricular" science communication activities may still be affected by participants' professional identities, as well as the collective nature of the institutions in which they participate as unpaid staff. Just as all of us think (intuitively) about the climate of public opinion, members of professional groups also think about the climate of professional opinion (although not, of course, by that name). How will their employers and professional colleagues react to their activities, both within and outside of the workplace? Some bloggers and freelance writers or producers who specialize in scientific topics may work for themselves, of course, but they are still dependent on science-related institutions as sources of information and inspiration, if not actual funding.[5] They are also part of the collective web.

All of these communicators very often belong to professional organizations, but even if they do not join them, they are still likely to be influenced by them as the "standard-setting" organizations for their professions. Just as the American Medical Association influences the practice of medicine, various journalistic organizations influence the practice of journalism, and both scientific journals and scientific societies (as well as major grant-making organizations, such as the National Science Foundation) influence the way science is conducted and the directions that it takes—that is, what science gets done.

The relationship between scientists and journalists is a highly symbiotic one. Some research suggests that scientists are increasingly sensitive to what the journalistic community thinks (and says) about their work, a phenomenon called "mediatization" (Weingart 2001). The converse is also true: Journalists (including science journalists) are naturally concerned about what their sources will think of their work—otherwise, next time around, those sources may not be so cooperative. The best journalists want to be independent, critical and cutting-edge, and many science journalists will not ask or allow scientists to approve their actual copy on principle, but they cannot claim to be entirely free of the need for approval, either. Like all journalists, they are dependent on their sources.

To be a professional—whether doctor, lawyer, teacher, journalist, or scientist—means to be a member of an organized group, sometimes one with formal standards and requirements. Professional norms, which can be explicitly stated in codes of ethics or other formal rules, or simply implicitly reflected in daily professional practice, are also the products of collective processes, reflecting the existence and practices of social institutions (such as media institutions and scientific or research institutions) and organizations (such as professional societies in both journalism and science). Understanding the professional actions and reactions of both journalists and scientists requires consideration of the organizational and inter-organizational context in which they work, which often determines what is considered acceptable and unacceptable behavior, what should be encouraged and what discouraged.

The social organization of professional science communication is today quite complex, with many overlapping groups. But the relationship between scientists and journalists remains an especially important one, and the professional norms and expectations that apply to these two groups (some shared, others unique to one or the other group) remain particularly important. Both groups serve as primary gatekeepers vis-à-vis the public at large for what science is considered valid and legitimate (as well as newsworthy) and how it will be framed and otherwise communicated to those outside science. In a very literal sense, scientists and journalists still "co-produce" the public character of scientific truth. In the remainder of this chapter, we first consider the historically derived professional norms and expectations of journalism and of science that are especially relevant to many of the challenges surrounding the communication of climate science in particular. Both journalists and scientists have embraced particular professional norms and expectations that have not always produced the best possible communication on climate change.

These two professional groups are not the only science communicators, by any means, as the above discussion has emphasized—and as the next chapter will elaborate. However, these two groups are both historically and at the present time arguably the most important ones, or at least among the most prominent and influential, and the relationship between these two groups has inspired several now-classic foundational studies of science communication (see, e.g., Friedman et al. 1986; Gregory and Miller 1998; and Nelkin 2005). But while scientists and journalists remain the most visible science communicators, with scientists communicating partly through journalism to the world outside science, things are changing. So in the following chapter we will then turn to a discussion of how contemporary developments involving a shifting landscape of media-based communication alternatives (including so-called "new media") and a proliferation of science communication actors and programs have changed the rules—or at least some of the practices and expectations—of science communication.

A Word about the Nature of Social Norms

Three overlapping sets of ideas govern whether human behavior is defined as acceptable: norms, ethical or moral principles, and laws. The word "norm" (or "normative") as used in social science refers rather generally to any social or cultural expectation about what people "should" do; a professional norm can reflect what is the typical or standard choice in a specific professional context, whether derived from long professional tradition (such as the admonition to "do no harm" or to "tell the truth") or developed more recently (such as an established "standard of care" in health and medicine). A norm can apply to expectations in any aspect of life, however (such as how someone should behave in a given social situation—or whether energy conservation should be seen as an obligation due to global warming). When norms determine what behavior should be considered acceptable as part of a system of distinguishing right and wrong, then for purposes of this discussion they are referred to as ethical norms.

Norms are not always expressed in formal written rules, and in some cases we may not even be consciously aware of them. Many laws reflect cultural norms, but not all norms are reflected in laws, so the two (formal laws and social norms) should not be confused with one another. And just as not all norms are expressed in written legal rules, not all of them directly concern ethics—that is, whether the behavior in question is classified as right or wrong. Something can also be unethical without being illegal. Dressing inappropriately for an important social occasion such as a wedding or a court appearance might break a norm but it wouldn't be illegal, and it usually wouldn't be considered unethical either.[6] And under some circumstances, something can be illegal (such as driving over the speed limit) but is socially tolerated even so. Our perceptions of others' normative and ethical expectations (another part of the climate of public opinion) and our observations of others' actual behaviors both influence us, and they have been incorporated into models of behavioral choice (in particular, the theory of planned behavior or TPB; see Ajzen 2012). However, this influence is not necessarily always easy to measure.

The norms of appropriate social behavior seem to shift perpetually, and issues of what is formally allowed and not allowed—the sort of rules that are incorporated into formal codes of ethics or even actual laws—shift as well, although perhaps not as rapidly. Some norms are routinely practiced but rarely articulated. When people act inappropriately, we often immediately recognize this even where they are not breaking a specific rule that we can articulate: We sense that something is "off" even when we can't say exactly what, in other words. Human beings seem to have an enormous capacity for interpolating and navigating such a complex set of rather flexible rules—ordinarily on an intuitive basis and without much thought. This is part of what makes us such a highly social species. It is very common for social norms, even ethical norms or legal rules, to be reinterpreted in light of a specific set of circumstances, and there are many gray areas that shift as well. This does not refer solely to attempts to rationalize bad acts, but to the many situations in which we have to pause and ask ourselves, "What is the right thing to do here?" Climate change has actually altered some of the answers, as we become more aware of the consequences of some of our actions, but this did not happen overnight and the process of coming to grips with climate change as an ethical dilemma remains ongoing.

Just as laws for personal behavior that have been formally adopted through legislative action generally reflect the existence of a consensus, specific expectations about professional practice that go beyond the law are often based on conscious deliberation and consensus at the collective level, and this consensus is expressed by organizations in the form of policies, practices, and codes of ethics. Violating consensus-based rules for behavior generally has consequences—in the legal system, prison terms and fines; in the professional world, in some cases expulsion from the profession (as when a lawyer is disbarred or a doctor loses his or her license to practice medicine), in others potentially the loss of a job or client and damage to one's reputation. Just as in the rest of society, professional groups also have many expectations that are taken for granted rather than formally deliberated and added to a written code. But at the same time many such groups, including journalists' professional organizations, do have written codes of ethics.

From a sociological perspective, ethical norms of right and wrong are understood to be constantly emerging rather than fixed. Yet expectations for social or professional behavior tend to have some level of continuity even in the face of change. Cyberspace provides a convenient contemporary example. "Netiquette", the code of behavior expected in Internet-based electronic communication, continues to evolve. This is largely an unwritten code; despite the fact that searching online for the rules of online behavior will turn up quite a number of lists of rules, few (if any) of these have been the subject of any formal deliberation or adoption. Most of these rules are still actively emerging, and there are rarely any immediate consequences from breaking them, beyond incidental social disapproval. Yet these behavioral rules were not made "from scratch", so to speak. Rather, the nature of virtual online communities and the behavioral expectations associated within them grew largely out of expectations prevalent in the broader culture (Jones 1995). What is acceptable in practice is continually revisited, yet broader ethical principles are often preserved.

The world of new media has influenced journalistic ethics as well. Is a blogger bound by the same ethical rules as a traditional journalist? What about legal rules—can that same blogger, or someone sending tweets, claim protection from the "shield laws" that apply to traditional journalists in many states? Some of these answers are still unclear.

Evolving Journalistic Ethics

Where did the concept of journalistic objectivity originally come from? In the United States, the eighteenth-century press was largely a partisan press, with specific publications often associated with particular political positions.[7] Nelkin (2005) traces a historical relationship (largely in the twentieth century) between objectivity as a standard in science and objectivity as a standard in journalism, but also claims that "journalists no longer believe that real objectivity is possible" and have substituted "balancing diverse points of view," fairness, and not conflating news with opinion as standards (p. 84). Other scholars have traced these concepts back further; for example, Schudson (1978) attributed the evolution of newspaper reporting styles from the partisan publications of the late eighteenth and early nineteenth century to today's more "neutral" approach to the rise of the Associated Press and other wire services, which essentially "sold" news (just as they do today) to publications and to editors in a variety of locations, with a variety of readers and varying political dynamics.

The current Society of Professional Journalists (SPJ) Code of Ethics[8] lists four overarching principles: to seek truth and report it, to minimize harm, to act independently, and to be accountable and transparent. Today, "balance" (in the sense of including both sides of every debate) is not explicitly listed among these, the oft-repeated claim by the Fox News channel to be "fair and balanced" notwithstanding.[9] The goal of being "balanced" is certainly a difficult standard to reach—perhaps at least as difficult as either "objectivity" or distilling absolute truth, although journalism students will still encounter the concept of "balance" in their classes and it is practiced by working journalists in many situations. However, in a country like the United States with just two major political parties, "balance" is often interpreted as representing *both* sides (that is, two of them: left and right) rather than *all* sides of an issue. This in itself can be a polarizing influence. This tendency likely reflects the nature of political reporting in a two-party system, but this interpretation of "balance"—which can reduce the most complex controversy to a two-sided debate—does not generally translate well to science.

The SPJ Code does ask journalists to "seek sources whose voices we seldom hear" and to "support the open and civil exchange of views", practices that it seems potentially reasonable to extend to minority views within science, at least under some circumstances. Reports of single-study results, for example, can appropriately be balanced by seeking an expert opinion from another scientist not connected to the study in question, adding context helpful to understanding the significance of the report. This is a reasonable and practical antidote to the tendency for the public relations staffers at universities (and other research-related institutions, including journals) to send out press releases treating every incremental development from within their institution as a "breakthrough".

Less easily justified is the balancing of a story about a controversial application of science (for example, in medicine or biotechnology) with a view from an ethicist or religious leader. While it is vitally important to include those views, this sometimes reinforces the impression that science and ethics (or science and religion) are inevitably at odds, without necessarily adding an alternative "balancing" perspective on the science itself. The use of "balance" as an implicit standard for achieving "objective" coverage is one of the constraints introduced by what Shoemaker and Reese (1995) called "journalistic routines".

While it is the nature of science that known truth will itself evolve and change, how best to report scientific dissent has remained a difficult problem (see Dearing 1995). In principle, dissenting views within science that have only minority support can turn out to be correct in the consensus judgment of scientists at a later date, whereas consensus views can turn out to be wrong (see Kuhn 1970). So a story about scientific findings, including climate change findings, sometimes gets "balanced" by the inclusion of a "skeptical" source, even where a strong scientific consensus exists. This is unfortunate, but it is an easy misstep to understand. It is not always ideologically motivated (although sometimes this might be the case, of course). The problem of "false balance" characterizes coverage of other issues, such as the relationship between autism and vaccination (Dixon and Clarke 2013). But achieving appropriate balance and fair reporting of scientific truth is certainly not easy, especially at the stage where a strong consensus within science itself does not yet exist. And once polarization has developed, as for climate change, walking it back seems very difficult.

At any moment in time the actual scientific truth surrounding a given topic or proposition can be still emerging or highly uncertain; in fact, it is most likely emerging or uncertain science that constitutes the news that people most want to read and that journalists generally seek to report, and there can be little doubt that journalism follows controversy. "Medialized" scientists can be expected to seek news coverage for their research programs and points of view, sometimes in the hope it will help their quest for funding. After the fact, we can look back and say that journalists should have known and reported the truth of whatever matter is at hand, but for science, this is not ever easy. Before there is clear consensus in the scientific community—which for some issues may take decades, even longer, and the consensus view is always subject to revision in the light of new evidence—it is expecting rather a lot for journalists, even science journalists, to be able to sort the wheat from the chaff on their own. And at one time, there was indeed less consensus within the scientific community about climate change than there is today.

Even now, many scientists remain reluctant to attribute any specific weather events to climate or climate change, sometimes leading to news coverage that can sound excessively tentative on this point. Scientists, as the next section will detail, are trained to rely on what the data indicate, and the data on the relationship between climate trends and today's weather is always a matter of probability, not "truth". Fortunately, despite all this complexity, progress has been reported for journalism in the United States, which may now be less likely to balance consensus-based scientific views on climate change with "maverick" skeptical positions than was the case just a few years previously (see Boykoff 2011). But this has been an uphill battle, and the "false balance" problem persists in some media.

Simply speaking out on climate may also be sooner said than done. Given the level of activity many observers see as being generated by organized "skeptical" groups, getting journalists to see and report the truth through the fog may be a bigger problem than simply reducing overreliance on the concept of balance. Shoemaker and Reese (1995), in their "hierarchy of influence" model, identify many of the influences that constrain the work of practicing journalists. These range from the influences of the background and training of the individual journalist to (as mentioned above) the organized routines that they learn to follow in performing their jobs; from organizational constraints from within their own media organization to those stemming from outside (governmental influences, advertising and public relations activities, and the various institutions and individuals used as news sources); and the taken-for-granted, culture-based, ideological context in which the work takes place. On a day-to-day basis, most of us find it difficult to act entirely independent of the social context that surrounds us—another "climate of public opinion" dynamic.

Today there are signs of change in professional views on the ethical norms of reporting science, however. According to Hiles and Hinnant (2014), leading environmental journalists still embrace the concept of objectivity but also increasingly advocate a "weight-of-evidence" approach as proposed by Dunwoody (2005), which calls for reporting of the direction suggested by most of the evidence and representative of the largest proportion of expert thought. Kohl et al. (2015) have studied reactions to journalism based on this advice and conclude that weight-of-evidence reporting may reduce some of the uncertainty that audiences perceive in cases where past reporting has been "lopsided". Differences of opinion can also be reported without leaving the impression that they are necessarily co-equal or equally accepted within the scientific community.

Unfortunately this begs the question of how journalists can independently deduce which evidence and which expert sources to trust—who is likely to be right and who likely to be wrong, who is ultimately credible and who is not. This requires a lot of time for research and a high level of understanding of how science works. It is often easier years down the road to look back and see what *should* have been reported as truth than it is to see it in the moment.

The National Association of Science Writers (2014) has its own code of ethics, which now includes the admonition that science writers "should support the healthy exchange of views and opinions in science, but also realize when scientific principles *are no longer questioned* by the majority of *reputable* scientists involved" (emphasis added). Undoubtedly this is responsive to controversies in areas such as climate change reporting. Yet, with the very best of intentions, even specialist journalists writing on tight deadlines and with limited resources will have difficulty keeping up with the state of the science on multiple issues. Furthermore, there is some suspicion that the proportion of journalistic science stories actually written by science specialists may be going down, another casualty of the economic pressure on today's news media.

When the science evidence itself is still uncertain, when many scientists may not yet be sure where the scientific consensus will end up, and when the idea that scientific views may be divided has permeated popular culture, journalists simply may not be in the best position to take sides and identify the (likely) truth, even if they are science specialists. The problem may be partly that uncertainty—like stigma—tends to be somewhat "sticky"; that is, it is persistent and seems to be hard to turn around even where new scientific consensus has formed, a point on which more research is needed. Various publics, for a variety of reasons, may cling to their beliefs that controversy exists—just as some do to this day on the subject of vaccination causing autism.

This points to another area of media influence needing further research, in some ways the complement of uncertainty: media legitimization. This is a likely candidate as a powerful media effect, and yet it is also one that is very much under-studied. Like other aspects of reputation, the perceived legitimacy or illegitimacy of points of view may "stick" persistently, and we can expect that any point of view that the media has legitimized by treating it as a part of legitimate science may have a very long life even after the consensus view has changed. It is not that the audiences for science are unintelligent, but most of us may not normally have much reason to pay deep attention to ongoing coverage of a particular topic. We may read about something once or twice, and then move on to newer topics. So, very often, do the journalists reporting on these topics.

Controversy should be reported, yet neither uncertainty nor legitimacy should be overstated. That is easy to say but unfortunately is something of a tall order in practice. Despite good training and good intentions, journalists are going to remain unable always to predict which version of "truth" will ultimately prevail. Scientists themselves cannot do this reliably. Arguably, the best science journalists navigate this territory with extraordinary dexterity, accurately reflecting both uncertainty and certainty, and incorporating seemingly legitimate but also minority or otherwise alternative views in an appropriately nuanced way. But there are not enough science journalists with this kind of acuity and skill in the world. And writing that kind of story takes a lot of time for anyone, which translates into a lot of resources—resources media organizations may not have at present.

In addition, alongside the emergence of the "dialogue" or "public engagement" movement in science communication more generally, our thinking about what science journalism ought to accomplish is evolving. Rather than merely filling gaps in public knowledge, in what has historically been thought of as the one-way transmission model, some scholars argue that science journalism can also seek to empower people and support greater democratization of science within society by, for example, including various stakeholder voices and more about the process of science (Secko et al. 2013). However, we are a long way from injecting this idealistic goal into everyday science journalism practice. Even if we could, as more and more people get their science news from a wide variety of Internet sources, they will be ingesting material written by all kinds of science communicators, professional and amateur, including many with purely strategic goals. This puts new demands on news consumers' critical thinking skills, which for the casual consumer of science news may be a challenge. This issue will be discussed in more depth in Chap. 6.

A New Ethical Landscape for Scientists?

A vital component of the social ecology of science communication is, quite naturally, scientists themselves (alongside the organizations they work for and belong to). Scientists have a largely well-deserved reputation for working hard and being intensely dedicated to that work. The traditional ethos of science documented many decades ago by sociologist Robert Merton, in work that is still widely cited, said little or nothing about communicating with non-scientists, however (see Turner 2007; Merton 1968; Merton and Storer 1976). The so-called Mertonian norms are communalism (knowledge and results are "owned" by the entire scientific community, which is engaged in a collective effort), universalism (scientific truth does not depend on where or by whom it is discovered), disinterestedness (science should be unbiased and not pursued for personal advantage), and organized skepticism (all scientific work is routinely and systematically subject to questioning and verification).

Merton's ideas were, of course, an idealized vision. In practice, these norms may or may not be fully embraced and may or may not be scrupulously followed. Scientists certainly have conflicts over the "ownership" of both ideas and data, for example, and it is fairly clear that work done by a recognized scholar at an institution with a strong reputation will likely be received differently from other work. Grant-seeking and industry employment can introduce serious issues of bias and advantage. The peer review process is not perfect and will not always screen out bad research. But the point here is less to decide whether these norms characterize science accurately, which is still a subject for debate, or even whether they are still applicable today, but to underscore the point that this most famous characterization of traditional scientific norms does not say anything about an obligation for scientists to engage with society more generally or to communicate outside the scientific community.

In terms of public communication, historically scientists have generally seemed to prefer a back seat. Scientists can certainly be key science communicators with respect to audiences outside science, in both formal and informal settings, and as part of their regular jobs or in addition to them, although it is not typically their main professional role. Yet the idea that scientists are under some specific obligation to do these things, let alone that they should be rewarded for them, is relatively new. Derivative works that teach—even college textbooks—are generally less prestigious as scholarly products than original research published in the peer-reviewed literature. And while a great deal of attention is given to ethical guidelines for the conduct of research and issues arising with publication of that research, much less is given to the ethics of communication about science to others.

Even for those willing to embrace the idea that scientists should be more regular communicators, it can constitute a tall order in practice. While many scientists are accustomed to communicating their results to other scientists, especially those within their own field, few have the skills or the time to reach out to other audiences both effectively and on a regular basis. Increasingly, though, they are under pressure to do better. The changing expectations for U.S. scientists along these lines (plus the unease of some scientists about their lack of appropriate communication skills) was documented over a decade ago by Mathews et al. (2005). Today, scientists in several countries, including the United States, state that they simply do not have the time to meet this obligation (Davies 2015). To the extent that "engagement in public engagement" is an emerging ethical norm for these scientists, it competes with other demands that are also seen as ethics-based, including the need to mentor advanced students appropriately and to do a good job in managing their labs.

What will it mean if public communication evolves into an ethical norm for all scientists? While we are still a long way from that, this would not necessarily be so radical as it might at first appear. The land grant university system in the United States[10] was designed to produce research useful to people outside the university, especially farmers. A significant investment in outreach activities and extension services were—and still are—part of the package. Yet even in the land grant colleges, science professors generally spend their time teaching and doing research, not giving practical instruction or advice to people who are not their students. Other staff members, often ones with less status and lower pay, take care of those obligations. Despite increased stress in National Science Foundation requirements for research grant applications to incorporate plans for "broader impact" activities (which have often been interpreted to imply more stress than in the past on outreach or dissemination), it seems likely that most scientists still spend most of their time in traditional ways. The time constraints that Davies documents may prevent many from deeper engagement with society.

Atmospheric scientist Judith Curry, among others, advocates for scientists to interact with those outside the ivory tower—including climate "skeptics"—via online social media means such as Twittering or blogging (see, e.g., Curry 2010). But this may not be the path for everyone either. The U.S. Bureau of Labor Statistics estimates there are over 700,000 scientists in the United States, excluding social scientists, technicians, and people classified as "postsecondary teachers" (U.S. Department of Labor 2015). If just *one in ten* of those scientists hosted a blog, that would be 70,000 of them. It is anyone's guess how many of those 70,000 blogs would have more than a very few readers, but it is clear that if maintaining an ongoing and active online presence ever evolves into an ethical norm for university-based scientists, the science "blogosphere" would quickly become rather crowded. Yet assuming that climate skeptics would be won over by a persuasive argument in a scientist's blog is probably unrealistic to begin with—a contemporary twist on deficit-model thinking.

The problem of "scaling up" the blog idea—while it is certainly an excellent option for some—has its own challenges. Science blogs raise new ethical issues of their own. What does blogging by scientists do to the usual prohibition not to discuss results from a paper or study not yet finished—or on which peer review has not yet been completed? Should these blogs discuss, even criticize, the work of other scientists? Or should scientists' blogs then be limited to discussions of established, widely replicated, findings? What should the scientific community do (if anything) about blogs that present viewpoints inconsistent with a strong scientific consensus, which will certainly continue to occur for climate? At least in a news story, views that are patently contrarian are generally presented in the context of more mainstream ones, an example of why judicious balance is sometimes still needed. For a "skeptic's" blog, achieving this is unlikely.

Fortunately, there are many other ways for scientists to become involved in public engagement and public communication—from simply serving as sources to journalists, as some have traditionally done, to participation in public activities with non-scientists at the local level (say, by discussing their work at a science café) to acting as advisors to government on emerging policy issues. Even so, this might come at a professional cost. In a phenomenon widely referred to as the "Carl Sagan effect" after the late astronomer who made regular television appearances during an era where this was even further from the norm than it is now, scientists who do become active and even well known for their engagement with society risk disapproval from their colleagues. If a scientist is spending time in public appearances, the thinking goes, then they cannot be dedicated to their work to the same degree, and if they are not that highly dedicated to their work, they are simply not good scientists. Only a few become the kind of "visible scientists" (Goodell 1977) who routinely make themselves available for media interviews.

Science is hard work and successful scientists do work hard, expecting their graduate students to do likewise. This means at least a full work week and likely many extra hours spent actually doing research rather than talking about it with others outside the university. The presumption is that a serious scientist *must* be completely dedicated to research, and if they are not, they may not be taken seriously as scientists. Even if this disapproval is not as strong as it once was, it remains a palpable influence.[11] Yet things are changing, and anecdotally the younger generation of scientists seems much more open to the idea that communicating to and interacting with the public, in whatever form, is in fact an emerging ethical norm associated with their work as scientists.

Yet the majority of scientists may still be reluctant to embrace a role in public communication of science. One recent study based on survey data from AAAS members argues that even though there is a manifest need for basic communication training among scientists, many scientists may actually be reluctant to receive this training (Besley et al. 2015). These scholars conclude that one of the strong predictors of willingness to participate in communication training is a belief that an ethical communication goal is being served; their respondents tended to believe that communication skills training aimed at improving their ability to explain science is ethical whereas training designed to make them better able to make audiences see them in a positive light is seen as less ethical. Least ethical of all: learning to use communication to place science itself in a more positive light. Interestingly, these survey respondents did not express much willingness to engage with the public online or through the media and only moderate willingness to engage face-to-face.

Thus a variety of factors—time constraints, skill constraints, a lack of interest, and a concern about ethics, plus the lack of an ethical imperative within the central traditions of science—limit the motivation of practicing scientists to become more engaged with those outside of science. What does this mean for climate science? Good science journalism is difficult and sometimes seems to be in short supply. Scientists themselves are unlikely to fill this gap anytime soon. We certainly need more participation by credible scientists in public discussion of climate issues, but it may not be realistic to expect this to evolve into an ethical norm for the average scientist, at least not in the short term. Further, for better or for worse, there are many other sources of science- and climate-related information available to almost everyone today on the Internet and elsewhere. The problem that comes with this is that information seekers may need new skills to recognize the relevant evidence and the most credible interpretation of that evidence. While the nature of public science communication is changing, it is already almost too late to avoid the worst effects of climate change. In this context, the next chapter will take a closer look at emerging trends in science communication as they relate to communicating climate change.

NOTES

1. That is, the Department of Energy, Environmental Protection Agency, Fish and Wildlife Service, National Aeronautics and Space Administration, National Oceanic and Atmospheric Administration, National Park Service, National Weather Service, United States Department of Agriculture.
2. Available at https://www.opr.ca.gov/s_listoforganizations.php
3. As yet, however, relatively few private nonprofit organizations seem to have been established to work specifically on climate issues. The National Center for Charitable Statistics has counted over 1.5 million different nonprofit organizations in the United States, based on tax records, but only 95 of

them have the word "climate" in their name (see nccs.urban.org). Of course, many organizations concerned with environmental issues more generally are today working partly on climate issues, and climate-oriented organizations may not always use names with the word "climate" in them, but this is still a revealing statistic.
4. However, the American Meteorological Society does certify broadcast meteorologists who hold what they consider an appropriate bachelor's degree.
5. Independent writers and other creators of science-related materials and programs may turn to private foundations for their funding, for example—or, rarely, to government grants.
6. It is, however, relatively easy to imagine times when inappropriate dress could cause social disruption and even create psychological distress for someone else. This could even constitute unprofessional behavior—such as an attorney showing up in gym shorts for court. Are such acts also unethical? Opinions could vary, but if the attorney's client lost her case a result, we would probably lean toward saying "yes".
7. This remains true in many parts of the world; consider, for example, the Italian press, which is divided along roughly the same lines as the country's complex political party structure.
8. Available at http://www.spj.org/ethicscode.asp
9. SPJ's code is not alone; dozens if not hundreds of other journalistic ethics codes exist, including the codes of individual media organizations.
10. Every individual state in the United States has at least one "land grant" university, so called because the federal government once set aside grants of land on which some of these campuses were originally built. In some states branches have multiplied to the point there is an entire network of "land grant" schools.
11. Such expectations matter. I can attest from personal experience in teaching a course for women in science (and women considering whether they wanted to be in science) that this requirement for total dedication to the lab has likely kept any number of women out of science. In recent years, I have been told about graduate students in science (of either gender) who felt that if they participated in "public engagement" activities their advisors might hold this against them.

References

Ajzen, I. 2012. The Theory of Planned Behavior. In *Handbook of Theories of Social Psychology*, eds. P. A. M. Lange, A. W. Kruglanski, and E. T. Higgins, vol. 1, 438–459. Sage.

Besley, J., A. Dudo, and M. Storksdieck. 2015. Scientists' Views About Communication Training. *Journal of Research in Science Teaching* 52(2): 199–200.

Boykoff, M. 2011. *Who Speaks for the Climate? Making Sense of Media Coverage of Climate Change*. Cambridge University Press.
Cobb, R. W., and C. D. Elder. 1983. *Participation in American Politics: The Dynamics of Agenda-Building*. 2nd ed. Johns Hopkins University Press.
Curry, D. 2010. Why Engage with Skeptics? http://judithcurry.com/2010/11/08/why-engage-with-skeptics/
Davies, S. 2015. Scientists' Duty to Communicate: Exploring Ethics, Public Communication, and Scientific Practice. Unpublished manuscript, University of Copenhagen, Denmark.
Dearing, J.W. 1995. Newspaper Coverage of Maverick Science: Creating Controversy Through Balance. *Public Understanding of Science* 4(4): 341–361.
Dixon, G., and C. Clarke. 2013. Heightening Uncertainty Around Certain Science: Media Coverage, False Balance, and the Autism-Vaccine Controversy. *Science Communication* 35(3): 358–382.
Dunwoody, S. 2005. Weight-of-Evidence Reporting: What Is It? Why Use It? *Nieman Reports* 54(4): 89–91.
Friedman, S. M., S. Dunwoody, and C. L. Rogers. 1986. *Scientists and Journalists: Reporting Science as News*. Free Press.
Goodell, R. 1977. *The Visible Scientists*. Little, Brown.
Gregory, J., and S. Miller. 1998. *Science in Public: Communication, Culture, and Credibility*. Plenum Press.
Hiles, S., and A. Hinnant. 2014. Climate Change in the Newsroom: Journalists' Evolving Standards of Objectivity When Covering Global Warming. *Science Communication* 36(4): 428–453.
Jones, S. G. 1995. *CyberSociety: Computer-Mediated Communication and Community*. Sage.
Kohl, P. A., S. Y. Kim, Y. Peng, H. Akin, E. J. Koh, A. Howell, and S. Dunwoody. 2015. The Influence of Weight-of-Evidence Strategies on Audience Perceptions of (Un)certainty When Media Cover Contested Science. *Public Understanding of Science*. Published online ahead of print, December.
Kuhn, T. S. 1970. *The Structure of Scientific Revolutions*. 2nd ed. University of Chicago Press.
Mathews, D.J., A. Kalfoglou, and K. Hudson 2005. Geneticists' Views on Science Policy Formation and Public Outreach. *American Journal of Medical Genetics Part A* 137(2): 161–169.
McCombs, M.E., and D.L. Shaw. 1972. The Agenda-Setting Function of Mass Media. *Public Opinion Quarterly* 36(2): 176–187.
Merton, R. K. 1968. *Social Theory and Social Structure*. Free Press.
Merton, R. K., and N. W. Storer 1976. *The Sociology of Science: Theoretical and Empirical Investigations*. University of Chicago Press.
National Association of Science Writers. 2014. Code of Ethics for Science Writers. https://www.nasw.org/code-ethics-science-writers

Nelkin, D. 2005. *Selling Science: How the Press Covers Science and Technology*. Rev. ed. W. H. Freeman.

Noelle-Neumann, E. 1993. *The Spiral of Silence: Our Social Skin*. 2nd ed. University of Chicago Press.

Revkin, A. C. 2006. Climate Expert Says NASA Tried to Silence Him. *New York Times*, January 26. http://www.nytimes.com/2006/01/29/science/earth/29climate.html?_r=0

Schudson, M. 1978. *Discovering the News: A Social History of American Newspapers*. Basic Books.

Secko, D.M., E. Amend, and T. Friday. 2013. Four Models of Science Journalism. *Journalism Practice* 7(1): 62–80.

Shoemaker, P. J., and S. D. Reese. 1995. *Mediating the Message: Theories of Influence on Mass Media Content*. 2nd ed. Longman.

Turner, S. 2007. Merton's "Norms" in Political and Intellectual Context. *Journal of Classical Sociology* 7(2): 161–178.

U.S. Department of Labor. 2015. Bureau of Labor Statistics. Labor Force Statistics from the Current Population Survey. http://www.bls.gov/cps/cpsaat11.htm

Weingart, P. 2001. *Die Stunde der Wahrheit? Zum Verhältnis der Wissenschaft zu Politik, Wirtschaft und Medien in der Wissensgesellschaft [A Moment of Truth? The question of Science's relation to Politics, Economy and Media in a Knowledge Society]*. Velbrück.

CHAPTER 5

Science Communication: New Frontiers

In this chapter, we look further at contemporary developments in science communication and its audiences, including those beyond traditional media, as these relate to climate change communication. People's knowledge of climate and climate science is being addressed on many fronts, and a solid majority now accepts the reality of climate change. But the work is not done. People do not always accept that the primary cause of these trends is human activity, and they may not be able to imagine what we can do to address this—circumstances that can create a sense of hopelessness and limit popular support and enthusiasm for needed policy solutions. Meanwhile, both journalists and scientists are embracing at least incremental shifts in professional practice in response to climate and other challenging issues. In journalism, the gradual shift is toward better and more evidence-based presentations of science stories rather than leaning toward false balance, and (at least in academic circles) the possibility of journalism designed to promote more public engagement is under discussion.

Yet financial and other constraints will continue to limit the time and resources, including skilled journalists, available to cover complex science-based stories in a meaningful way. (The same can, of course, be said for coverage of politics, economics, international relations, and complex social issues reporting more generally.) At the same time, while many environmental groups support policy change to address climate, a non-governmental (NGO) infrastructure focused first and foremost on climate is still developing. Journalism routinely responds to active advocacy voices, which provide another important source of information subsidies.

In science, the hoped-for shift concerns in part how scientists see their role vis-à-vis the rest of society, with no existing tradition of an ethical imperative for public communication to guide them, not always much support from colleagues or administrators, and often limited self-confidence and little training in popular science communication skills and strategies. The possibility of career damage (or at least fears of career damage) for scientists who make a priority out of public communication is difficult to rule out, and powerful time and energy constraints operate here as well. Scientists are already busy people, and just like media institutions, universities (especially state-funded ones) are often having to do more with less.

Today there is considerable discussion of the need for scientists to engage more with those outside science, an idea that receives encouragement from a range of academic voices and from some major scientific societies (including the American Association for the Advancement of Science, AAAS). Major change in the role of scientists may not be coming quickly, and it may not be realistic to expect scientists themselves to take up most of the slack in public communication and public advocacy on climate issues—despite the fact that a few very vocal and visible scientists like former NASA scientist James Hansen and the late Stanford biology professor Stephen Schneider (among others) have been important leaders in directing public attention to climate. Direct advocacy may not always be for everyone, but everyone can make a contribution.

Anecdotal observations would seem to suggest that a higher proportion of younger scientists are increasingly interested in public engagement activities, suggesting that new generations may be more ready to embrace this idea. However, comparison between 2009 and 2014 Pew Research Center polls of AAAS members[1] on this point (Rainie et al. 2015) shows little or no real change overall. In both 2009 and 2014, just 2 % of these members reported they "often" write for science blogs, and in both years just 3 % reported they "often" talk to reporters. The number of those who use social media to talk about science in 2014 (not asked in 2009) shows that this varies with age, perhaps not surprisingly: Of scientists under age 35, 70 % did this "often" or "occasionally", whereas the corresponding number for those 65 and older was 30 %. But are they really speaking out more than previous generations, or just using different media? Even though programs to encourage public engagement with science should be seen as a positive and progressive trend, a major shift in the way most university-based scientists spend their time seems not to be very realistic anytime soon.

These findings tend to beg the question of what "counts" as public engagement. Most communication scholars and other social scientists interested in developing alternative public engagement forums—those advocating for a shift from "deficit" to "dialogue" and "participation", in other words—often seem to have in mind face-to-face interactions, either among citizens or between citizens and scientists, at least as an ideal type. Using social media (which is often used for one-way rather than truly two-way discussions; Lee and VanDyke 2015) and talking to reporters (with traditional media being the classic example of one-way communication) remain valuable forms of public communication, but they are not necessarily the type of thing scholars have in mind when they talk about public engagement, as opposed to a more clearly two-way process involving both scientists and non-scientists. Can this be done just as effectively using communication technology instead of physical presence? This is another area where more research is needed.

Social media can certainly be used to augment face-to-face discussion; however, some evidence suggests that deliberation participants prefer face-to-face discussion, based on the experience of a National Citizens' Technology Forum trial program on nanotechnology. Of the 74 citizen participants at six different university sites across the United States, only 3 % stated a preference for online participation after their forum discussions were completed, while 70 % preferred face-to-face discussions and the remaining 27 % did not prefer one over the other (Hamlett et al. 2008: p. 10). Interestingly, the experience had made a big difference: Going into these events, six times as many (18 %) had said they preferred online communication and just 33 % had said they preferred face-to-face discussion. While comfort with new media may have increased since 2008, it is worth paying attention to the fact that with more experience, these participants tended to be less positive about online (as opposed to in-person) discussions.

In one common vision of how this idea of citizen deliberation should work, non-scientist citizens would have opportunities to share their own knowledge and perspectives with scientists, as well as scientists having the opportunity to communicate to the citizens. The playing field would be more or less equal, in other words, without disrespecting either the citizens' values and goals or the scientists' unique expertise. This responds to concerns that scientists often fail to understand public perspectives (see, e.g., the classic case study of post-Chernobyl sheep farming by Wynne

1989, which famously illustrated how both scientific and lay expertise have value). Scientists could learn from citizens, not just the other way around, by considering value based and experience-based, as well as evidence-based, arguments.

Opportunities for broader public engagement with science are gradually growing as a complement to scientists' and their institutions' largely one-way outreach efforts directed at others outside the university. Increasingly, in line with the science communication field's present emphasis on discussion and dialogue, some science communicators working in academic settings, in science museums and centers, and occasionally in NGOs or government are actively exploring ways to expand opportunities for non-scientists to interact with scientists (whether through science cafés, science festivals, interactive exhibits, demonstrations, new media, "consensus conferences" and other deliberation, or actual participation in "citizen science" projects, where non-experts help to gather and analyze scientific data). Organized public deliberation forums generally seek to provide uniquely two-way opportunities for expert-nonexpert interaction. In different variations on this theme, the exact role of the scientist can range from discussion or activity leader to "sideline" expert advisor who leaves the actual deliberation to non-scientist participants. In some versions, people can engage in discussions of science-related issues without scientists themselves even being involved, except possibly to help create or vet informational materials. Thus the citizen-participants and their arguments take center stage.

But whether any of these approaches can be "scaled up" from the level of academic experiment to an effective and influential national and international forum for public discussion of science-related issues, including climate change, is something that has not fully been tested. Indeed, one complaint sometimes registered by participants in such projects is that their conclusions do not seem to get much attention from those within official policy circles. It is one thing to suggest we continue to experiment with these new formats, and quite another to say we have achieved any fundamental shift in the way science policy is made or science itself is governed. And, as the previous chapter concluded, even if these activities were scaled up, it may not be realistic to expect all scientists to participate.

Larger-scale events have been tried. In one of the largest such efforts, which was directly concerned with climate issues, the World Wide Views organization chose "climate and energy" as the theme for its third major deliberation event, which took place in June of 2015 (leading up to the

COP 21 United Nations Framework Convention on Climate Change meeting in Paris that December). The organization, a global effort to stimulate citizen deliberation through locally organized participation, reports it was able to engage nearly 10,000 citizens in roughly simultaneous, carefully orchestrated, public deliberations taking place in 76 different countries on June 6. The group presented its results and citizen recommendations in a series of "side meetings" at the COP meeting (see climateandenergy. wwviews.org for details). This was an impressive effort. Whether this had any major effect on the COP 21 deliberations, which produced what is arguably the most important international consensus to date on the need to reduce carbon emissions, is difficult to ascertain however.

At a minimum, those involved in organizing the World Wide Views deliberations did provide information on climate and the opportunity to consider policy alternatives in light of that information to many thousands of citizen participants around the world. Future World Wide Views activities are contingent on finding adequate funding, however, meaning that a continuous stream of similar future activity cannot be guaranteed. So, on the one hand, this effort demonstrated that larger-scale engagement programs can be made to work; on the other, their impact has yet to be documented, and the organizational infrastructure needed to support these activities on an ongoing basis is (at best) still emerging.

Having said that, these new approaches do help illustrate how a serious consideration of the present science communication landscape as it is relevant to climate change communication needs to go well beyond journalists, scientists, and the relationship between these two communities—vital as those relationships may be. Smaller deliberative events may face lower logistical challenges, including the kinds of public meetings often used for local-level decisions, meetings that tend to consider concrete and immediate choices rather than abstract national or global ones of such stunning complexity—both political and scientific. While all of these new initiatives seem nevertheless to hold some promise as ways to break down barriers between scientists and members of more general publics, at least one scholar has concluded that experiments with deliberative democracy for science consistently fall short of organizers' goals (Scheufele 2011). Some "public engagement" efforts seem more like public lectures, with scientists reverting to their more traditional academic roles. Even these seem to have value, however, and some evidence from the United Kingdom suggests adult participants often prefer lecture formats to more dialogic science festival options (Fogg-Rogers et al. 2015).

Regardless of longer-term prospects for these approaches, climate change poses a particular challenge, not only because of the scale and abstract nature of the problem and the speed with which we must begin to develop viable solutions, but because of the existing attitudinal polarization. While success with organized forums designed for deliberation seems somewhat mixed, and with the ideological diversity and geographic range of the United States making it very difficult for a small sample of participants in such events to "stand in for" the public at large, we should certainly continue to explore them. But we should also be realistic about what they will accomplish, especially in the near term. In a country in which many people do not even bother to vote, broad public participation in sustained consideration of climate science and what it means for energy policy will be difficult to achieve.

The archetypical professional science communicator may be a print journalist, just as the archetypical professional scientist is employed by a university and works out of a laboratory, but many, many other professional communicators also deal with science: corporate public relations practitioners; public information staff working for agencies of government; university news service staff and outreach or extension workers; advocates, including fund-raisers, who work for non-profit organizations; science center, science museum, and natural history museum staff; broadcast, film and video producers and scriptwriters; web content writers and producers and website creators; advertisers and marketers for "green" businesses and technologies (and for others who want us to think of them this way); and even actors and television personalities. "Amateur" science communication volunteers who are not necessarily scientists or professional communicators make science-related YouTube segments, volunteer with conservation organizations, write blogs about science, and find many other ways to participate in communicating science.

All of these groups and more put out the information that helps shape public perceptions of science generally and climate issues in particular. Beyond scientists and journalists, in the remainder of this chapter we will consider some of the other centrally important ways in which the science communication landscape—that is, the social ecology of science communication, as well as science communication itself—is changing. A great variety of voices and actors are making use of new opportunities to speak directly to audiences about science—including climate science. Often, these opportunities are connected to the availability of new technologies. These "democratizing" forms of communication are available to everyone—not just science communicators, professional or volunteer, but also to both climate advocates and climate "skeptics", presenting new challenges as well as opportunities.

The New Knowledge Brokers: New Media, New Actors

The Internet, which of course is not going away, is where we all tend to turn for information, in particular information about the latest developments in science or medicine, and in many cases for discussions and interpretations of these developments as well. This technology has helped fill much of the gap left by the economic restructuring of traditional news media—although, of course, many observers feel this same technology also helped to create that very gap by disrupting the economy of the news business, providing a seemingly infinite amount of informational material largely free of cost to the end user. Along with the explosion of scientific information itself, this migration to the Internet has helped change the structure and context of science communication in important ways.

Meanwhile, while more scientific research is being done and more scientific journal publications about it are being produced than in previous generations, the traditional news media still seem to be shrinking, not expanding, and their coverage of science (outside of a few major outlets) seems to be shrinking fastest of all. Those news organizations who have given up producing print news and migrated entirely to the Internet—or who try to maintain both presences—continue to face a huge challenge in making this shift financially viable. Start-up web-only news sources are proliferating, but it's not clear whether they are financially viable either.

In short, the expanding niche that is scientific discovery (and, occasionally, scientific controversy) continues to grow while the media's ability to report on science, as well as on all the other news of the day, is under sustained financial stress. How this will all settle out for the news business generally is well outside the scope of this discussion. But what also matters in our context is that this expanded niche is occupied by new types of knowledge brokers, that is, additional groups of people "whose job it is to move knowledge around and create connections between researchers and their various audiences" (Meyer 2010: p. 118). When actual biological ecologies are upset, whether by pollution or climate change or habitat loss or incursions by invasive species, stable niches get disrupted and there is new competition for the existing space, often resulting in increased species diversity—at least in the short term. The same is true of the social ecology of science communication: New media are in a proliferation phase, but a simpler constellation of successful organizations and channels seems likely to emerge in future.

As traditional print journalism has become less dominant, new actors have gotten involved, new media have emerged, and some existing actors and media have become more important. In the United States, cable news sources, for example, have greatly expanded over the past several decades, taking audience share from traditional broadcast efforts and possibly creating new audiences as well, and while these new networks may not carry a huge proportion of science-related material (given that broadcast-based science-specialist journalists, other than weathercasters, are not visible in plentiful numbers), it does reach most people in the country: 83 % of U.S. households subscribe to paid cable television (James 2015). Traditional broadcast news, itself also largely delivered today by cable or other broadband system rather than actually over the airwaves, remains a main news source for many—but they are not necessarily the ideal medium for science coverage since only a limited number of stories of modest length can be covered on a given day.

Cable television also poses its own problems. The proportion of time devoted to pseudo-scientific activity on non-news cable channels, from searching for "bigfoot" to contacting people's dead relatives, is disconcerting. Various forms of reality television—popular and cheap to produce—predominate, and they often blur the line between fact and fantasy. Cable news outlets that seek audience share by adopting specific political slants (MSNBC, Fox) should be good for democracy, so long as the facts are respected, but likely contribute to audience fragmentation along political and ideological lines—and so far, network perspectives on climate can seem as ideologically driven as the perspectives of the individual politicians quoted.

Should any reader of this book doubt that traditional journalism is on the wane, consider the countless newspapers that have gone out of business or gone to online-only publication in recent decades. Online revenues have not risen fast enough to make up for drops in the revenue that traditionally came from print-based advertising (see Barthel 2015). One study estimated that as many as one in five journalists working in 2001 was gone by 2009, with many more layoffs foreseen (Saba 2009). Of course, some people will turn to other media (centrally broadcast or cable media) for their information, but these media (aside from specialized science programming, likely reaching equally specialized audiences) devote little time to science. For this, many people will turn to the Internet—where the full range of knowledge brokers seek to attract audiences in new ways and where answers to almost any question can seemingly be obtained with a few keystrokes. Local newspapers often survive, but rarely have the resources or expertise to consider complex science stories. Indeed, climate policy activists might do well to target such publications with hyper-local climate information.

The niche that traditional journalism occupied in our social ecology of science communication is now being colonized by this more diverse range of actors. The downside is that traditional journalists operating in traditional, so-called "legacy" media served as important and trusted gatekeepers for science-related information; the remaining ones still do, to a large extent, although by their sheer numbers they may no longer be the dominant actors in this role. What science coverage remains seems more likely than ever to reach primarily audiences of the already interested, and science-literate, public. Journalists of all types make informed judgments (even if sometimes imperfect ones) on behalf of the rest of us about which scientific opinions operate within the sphere of consensus-based legitimacy and which are from the crackpot fringe, and the best of these reporters seek to help us understand "what it all means" as well. But their influence today, along with that of traditional broadcast anchors, seems greatly diluted.

In short, now that a wide range of actors can speak directly to audiences via Internet technology and with the journalistic community seemingly continuing to shrink, their gatekeeping and interpretive functions both appear weakened as well. A wider range of knowledge brokers are active, and the difference between legitimate and fringe positions may be harder to discern for the casual Internet browser—as well as, apparently, for some of these brokers. Considered optimistically, having more diverse news sources—including sources reporting from different political perspectives—really sounds like better democracy, but even so this shifts the heavy burden of recognizing legitimate from illegitimate claims largely off of the shoulders of news producers and onto those of information consumers. Many of these consumers may be ill-equipped to make these distinctions for scientific claims; most of them have other day-to-day priorities.

Broadcast/cable and remaining print media news sources certainly still offer some excellent science reporting, but there are built-in limitations. Broadcast coverage generally needs to be quite brief (and although some 24/7 cable news outlets fill some of this ample time with extended panel discussions, they are not very often about science). And longer-form reporting in major newspapers and magazines generally reaches elite audiences especially interested in science. The same can be said of the small number of excellent broadcast programs being made specifically about science. This may work well for those of us who fall in that "especially interested" category, but for the rest of the population good information about evolving science may be encountered only rarely, especially for those who aren't actively looking for it. It may be hard to locate—and it may be quite difficult to evaluate. Sound bite coverage of climate change, whether concerning the "hockey stick" graph from *An Inconvenient Truth* or the "climategate" emails, may be little more informative than those ubiquitous pictures of stranded polar bears.

Even those actively seeking climate information may have difficulty. Google "climate change" or "is climate change real", and you will readily find any number of choices—some accepting this reality, some disputing it, and some seemingly neutral (but look more carefully because that is sometimes a façade for "skeptical" views). Some (maybe three or four of the first ten alternatives on any given day) will be from the online versions of traditional media like magazines or newspapers; a few might be from online-only publications. Some will probably be from corporate sources, boasting about their existing clean energy contributions if not actually denying the need to go in a different direction. A few will probably be from governmental agencies or NGOs. And some, of course, will be from out-and-out climate "skeptics".

This could be described as a healthy and democratic "marketplace of ideas" from which information consumers can pick and choose, or it can be thought of as a skewed collection of material representing primarily special interests and heavily supported by "information subsidies" (products of the subsidized creation and distribution of information and reports intended to shift the news media agenda and therefore influence public opinion; Gandy 1982). The problem is that it isn't always easy to tell the difference between a self-serving message and an educational one. "Greenwashing" by corporate messengers is common. Disguising "skeptic" perspectives as mainstream climate science is also common. As a result, much depends on the skills of those information consumers to identify the voice of reason, an issue which the next chapter will address in much more depth.

Yet very few of us, *including those with scientific training*, can entirely make sense of the original research literature published in academic journals on most scientific topics unless we have particularly followed that topic, understand its broader context (for example, the existence and legitimacy of competing ideas), and have some notion of the credibility of whoever is involved and of their institutions, of the scientific papers that preceded it on the same subject, of the strengths and limitations of its methods, and of its particular vocabulary and assumptions—not to mention the current tone of discussions on the subject that are taking place at scientific meetings and in university hallways. We are all dependent on news reports and other forms of knowledge brokerage in many areas; one could argue this is especially true for science. As the scientific literature continues to explode in volume, and the work of individual scientists becomes more specialized, this isn't going to get any easier. We all need knowledge brokers. But many, perhaps even most, of today's most visible Internet knowledge brokers represent particular interests, from conservation organizations to energy corporations (traditional, "greenwashed", or actually "green") and from universities to policy think tanks (whether conservative or liberal).

Why is this such a problem? Having a diversity of voices should balance out the marketplace, one hopes. The existence of information subsidies may not always preclude the presentation of consensus positions or result in a biased picture of available and emerging scientific knowledge. However, it always opens the door to that possibility. The process through which most science communication took place even a decade or two ago, which we can think of as largely well-described by the old linear "transmission model" of mass communication, incorporated plenty of subsidized information. However, we could generally make the assumption that professional journalists served as the gatekeepers, interpreters, and even shapers of the information they received from press releases and institutional sources. (This system was never perfect, of course.) It is in this context that the practice of seeking alternative views on the value of particular scientific developments became firmly established. Done well, this is appropriate. Done superficially, it opens the door to subsidies from groups (including, for climate change, both true climate "skeptics" and anti-regulation voices representing the interests of particular segments of the corporate world, such as fossil fuel-based energy producers) who are not necessarily speaking in the interests of scientific truth. This is exactly the dynamics that are causing both scholars and practitioners to rethink the concepts of "objectivity" and "balance" for science and environmental reporting.

A small clarification is in order: Scientists routinely complain that media reporting is not focused on scientific truth to begin with, so the situation described here may strike them as an old story rather than a new challenge.

But in such cases those scientists seem most commonly to be thinking in terms of completeness rather than factual accuracy. In a journal article, where they are accustomed to publishing, every detail of method needs to be reported, along with every important limitation of the study. News reports need to be readable and interesting to news audiences, and methodological details normally get sacrificed (with, to be sure, implications for accurately conveying levels of uncertainty). The implications of the work may be overstated in some cases, again in an attempt to appeal to news audiences—or understated in others. These are not, however, the same as reporting things that are generally recognized by the scientific community to be simply wrong. A simple google search will generally turn up many things that are generally recognized as wrong by topical experts. It is up to the searcher to determine.

Universities and affiliated research organizations often do an excellent job, both within and outside of traditional journalism, brokering the knowledge generated by their own research staffs to outside audiences such as policy-makers, funding agencies, donors and potential donors, current and future students and faculty (and their families), and other friends of the university. The traditional press release (albeit in electronic form) continues to be a staple format, but the university's own websites and email lists are being put to work here as well. This often provides a genuine service to the community outside the university, backed by competent staff committed to that goal, but at the same time it is not entirely disinterested or altruistic. Universities, even those that are non-profit and state-supported, thrive or falter depending on their reputations just as much or more as do for-profit corporations. Their press releases are a form of information subsidy, in other words, just like corporate press releases. Nevertheless, there are a few brakes on this process: If a central goal is reputation enhancement, alongside information dissemination, inaccuracy is a big risk to take.

As an example of what can happen when even state-run universities work too hard to promote their own research, many readers of this book will remember the "cold fusion" controversy at the University of Utah, where in 1989 researchers Stanley Pons and Martin Fleischmann reported their radical energy "discovery" (later largely discredited) and made a public pitch about it at a university press conference. Their work, which had not yet been peer-reviewed, was represented as a radical nuclear breakthrough that held promise for producing abundant energy, basically for free. Their claims were premature, yet generated great global interest, and both they and their university were seeking research funding for this effort from a variety of sources, including the U.S. Congress. By taking the radical step of bypassing traditional gatekeepers—particularly journal peer review, in this case—to "go public" with early (and, as it turns out, misleading) data, no doubt they thought things might move faster.

Were Pons and Fleischmann aware that their preliminary findings were never going to pass serious scrutiny and would not stand up to replication attempts? That remains uncertain, but sometimes even researchers are not the best judges of their own work. Historian of science communication Bruce Lewenstein (1995) argues that the sequence of events can best be understood as reflecting the existence of many new communication channels creating new forms of research dissemination, including the ability to fax papers and preprints around the globe and the rise of Internet-based interpersonal communication (at that time, often in the form of electronic bulletin board and "newsgroup" posts), spreading information in a speedier and far more complex way than the one-way "linear transmission" model would ever suggest. There was a great deal of information in circulation and widespread uncertainty among journalists and even within science itself. The world of science communication had already begun to move much faster—and far less predictably. And the sheer volume and travel speed of that information apparently helped to confuse a lot of people. Today, this trend affects all of us, not just journalists.

German social scientist Peter Weingart (introduced in the previous chapter) claims that the entire university research enterprise has begun to adapt its priorities to media demands for news in a process he terms "medialization" (Weingart 2001). In retrospect, we can thus see the cold fusion controversy as a forerunner of times to come in which the first-tier marketplace of scientific ideas has moved from the "inner circle" of the scientific community and science-specialist journalists to the broader society, social media is increasingly a medium of choice for young scientists to discuss research, and universities invest significant effort and resources in promoting their own research.

In a related development in 2009, a group of research universities that now includes almost 60 partners, primarily in the United States, banded together to produce the non-profit research news site Futurity (www.futurity.org), supported by a consortium that distributes its news directly to the public, free of charge. Lisa Lapin, Stanford University's assistant vice president for communication at the time and a co-founder of the program, described the move as creating "a direct link to the research pipeline" that was initiated "in an era of shrinking traditional news media" (Orenstein 2009). This may well have been a brilliant move: Futurity news articles seem solid, are written in interesting and accessible prose, and gain credibility by including links to original sources in the peer-reviewed research literature. And Futurity contains numerous articles relating to climate (1185 of them in a recent search), including some analyzing mitigation and adaptation strategies.

This website reports primarily on specific findings from individual studies or particular researchers or teams, the type of material historically disseminated by universities through ordinary press releases (and inviting criticism when journalists used that material unaltered as the exclusive basis of news stories). The site provides a means to showcase and promote the university-based research of its partners to interested outside audiences—not necessarily to identify broader trends, consider competing interpretations, or provide extensive context. They reflect the interests of a fairly small number of participating universities (roughly just two percent of the nearly 3000 four-year degree-granting institutions in the United States; National Center for Education Statistics 2015). One has to wonder how many readers visit the site and who they are; regardless of the answer, and despite the generally high quality of the material, this "information subsidy" approach is not a comprehensive substitute for more and better science journalism. Communication researchers might consider better documenting the implications of this trend, especially in a world where more for-profit universities are appearing—and in which all universities need multiple sources of funding to survive.

Beyond the Ivory Tower

Meanwhile, outside of universities and government agencies, myriad other non-news organizations have also been moving further into the "knowledge brokering" business for science, with advocacy voices ranging from various corporate and commercial interests to a variety of non-profit groups—all promoting their own agendas. Their direct route to the public is the Internet, without even a Futurity-like site to serve as intermediary.

Not surprisingly, conservative political interests opposed to extended governmental regulation on principle will also be opposed to further regulating the energy industry; liberals, progressives, environmentalists and pragmatists who want the government to intervene will feel otherwise, of course. The polarization here can be read as reflecting "the tension between those defending the current economic system and those willing to acknowledge environmental degradation as a consequence of industrial capitalism" (McWright and Dunlap 2011: p. 157). As they point out, Al Gore's original climate movie *An Inconvenient Truth* made the ("deficit-model") assumption that informing people about the science would be sufficient to persuade them. Conventional wisdom, in hindsight, suggests that if people didn't like former presidential candidate and liberal Al Gore, they tended not to accept the climate change message.

Science communication scholars, aware of the failure of this kind of "deficit" thinking, should not have been at all surprised that this premise turned out to be simply wrong. We should take the lead in training future generations more broadly on this point. Scientific controversies in the public sphere involve political and economic interests, social values, and issues of trust. Although scientific facts get used as ammunition in such cases, social controversies about science rarely revolve around scientific fact exclusively. Awareness of the complex relationships between political interests and scientific arguments should be addressed as part of basic science education.

Nor should we be surprised that energy interests have their own message strategies. As an example, a television advertisement that appeared widely on political broadcasts in the United States during the 2016 election season portrays "energy voters" as those in favor of development of "all" of America's oil and gas reserves; the sponsoring organization, Vote4Energy.org, offers an "energy voter" pledge on its website that asks people to agree to support this development, to support policy that "leverages America's full range of energy resources," and to get behind other "forward-thinking energy policies that sustain and grow America's prosperity" and its global influence. Sound reasoned, optimistic, and neutral? The group also warns on its site that this "promise ... can be wasted with restrictive and stale policies." While it is difficult to find this out on the Vote4Energy website itself, the *Huffington Post* (among others) attributed the campaign to the American Petroleum Institute, which Greenpeace has called "Big Oil's top lobbying group" (Gerken 2012).

Some conservation and environmental organizations have certainly placed addressing climate change near the forefront of their agendas, but the polarized political climate in the United States complicates this work greatly. Attempts to reduce the extraction and consumption of "dirty" sources of energy in favor of reliance on "cleaner" ones—for example, attempts to reduce coal use—raise concerns about job loss, especially in "coal country" areas where coal mining has supported generations of miners fearful of their livelihoods disappearing. This can also put a traditional liberal constituency, blue collar labor, in the same court as anti-regulation conservatives—thereby upsetting "politics as usual" in the United States.

The idea of population control, which is also sometimes suggested as a strategy for reducing the production of greenhouse gases along with many other forms of environmental degradation, has never been popular in the United States. Its promotion is likely to run up against opposition from both political conservatives and some religious groups. Rational as it may seem to be, it won't go very far in a political and social climate in which the area of reproductive choice is a hot-button political issue associated with abortion and with the stigma of government control over personal reproductive rights. This is likely true even though a 2012 Gallup poll showed that 89 % of Americans, including 82 % of Catholics, feel birth control is "morally acceptable" (Newport 2012).

While it is beyond the scope of this discussion to consider all of the complex political dynamics that bear on support for climate-related policies, creating a broad public mandate on addressing climate will require that traditional environmental organizations find ways to craft climate-related activities and messages in ways that appeal to those organizations' traditional constituencies—and perhaps, in the process, attract new ones. But that is not easy either. Existing environmental and conservation organizations that were established before the magnitude of the climate issue became clear generally have historical constituencies with a narrower focus. For climate change, positions and statements often reflect these organizations' pre-existing missions. For example, wildlife and habitat protection organizations focus on pleas to help threatened animals, including (for climate) those iconic polar bears. This works for the organizations' traditional constituencies, but in itself is unlikely to expand these organizations' appeal to the kind of broad constituency that is needed to fully address climate. For that, it seems likely that we will need to create new organizations. Indeed, this is already happening, but perhaps not quickly enough.

Consider, as examples, problems encountered by just two of the better-known environmental groups in the United States, the Environmental Defense Fund (EDF) and the Sierra Club, each of which has adopted broad mandates on climate. EDF (www.edf.org), which sees its mission as harnessing both market forces and science in the interests of environmental goals, prides itself on moving beyond traditional constituencies—for example, by partnering with businesses and, according to news reports, funding politicians like Republican South Carolina Republican Congressman Lindsey Graham, a conservative nuclear power proponent who believes in climate change (see Adler 2014). Graham had participated in the early crafting of a (failed) U.S. Senate cap-and-trade bill, to which he had reportedly added extensive subsidies for the nuclear power industry. Yet the environmentalist community has been a major enemy of nuclear power. EDF has also been widely criticized for getting too close to big energy interests in its position on fracking (Song and Bagley 2015). We do not claim that EDF is necessarily insincere in its stated positions, only that it may find the policies that it considers environmentally wise (from its pro-market, pro-science beginning point) might not be very popular with its own traditional constituency. For better or for worse, many segments of the environmentalist community to which EDF is apparently trying to appeal will likely be uncomfortable with either of these moves.

Oddly, there are polar bears on the EDF donation page but scant mention of nuclear power anywhere on the site—and little visible discussion of any current EDF nuclear position, although they do defend their position on fracking, which calls for strong regulation rather than a moratorium. Interestingly, in December of 2015, EDF launched a fact-checking initiative aimed at improving science journalism as well—an expression of their stated commitment to science-based policy. In the end, however, people's positions on energy issues will reflect issues of trust and values just as much as they do the science itself. This doesn't mean that getting the science right is unimportant, but getting all the scientific facts straight will never be enough to persuade everyone in most contexts. Even if the science should suggest that nuclear power and fracking may be better energy options than coal, at least in the shorter term, they are going to remain very difficult to sell to environmentalists.

Another well-known environmental voice, the Sierra Club (www.sierraclub.com), which describes itself as "the nation's largest and most influential grassroots environmental organization", expresses a clear commitment to clean energy, as well as a clear nuclear power policy and a clear fracking policy—they simply say "no" to both. Founded by California naturalist and early wilderness trekker John Muir, the Sierra Club historically emphasized the recreational and scenic alongside the educational or scientific value of wilderness. Their website tagline says we should "explore, enjoy, and protect the planet", in that order, while its banner admonishes us to "move beyond fossil fuels, preserve our wild America, and enjoy the outdoors"—in *that* order.

But they also provide another case in point, from their slightly different place on the spectrum of environmental politics: In trying to appeal to both their historical constituency of wilderness users and a more contemporary one of clean-energy advocates concerned about climate, two groups that likely overlap but are certainly not the some, some tensions are apparent: According to the site, "development of wind raises strong concerns among Sierra Club members" and the Club "opposes development in protected areas", including any area with "special scenic, natural or environmental value" (Sierra Club n.d.). That is a lot of areas. One hesitates to propose that wind generation should always be promoted regardless of location; nevertheless, it seems likely that an organization with climate as its central mission would not hedge its support for wind power quite so extensively.

The above two examples were chosen because both organizations are generally well-known and widely respected (though both have been the subject of criticism as well); this discussion is not intended to suggest either one is or is not doing good work. In general, it is not our intent in this book to criticize or to praise any particular environmental or climate group's approaches or policies. But these examples serve to highlight how tensions can arise in reconciling the historical missions and constituencies of existing NGOs that have a general environmental mission with the urgent necessity of addressing climate—and also to illustrate that leading environmental groups can reach fundamentally conflicting positions on energy issues, rather than speaking with one voice.

We can be confident that over time, more organizations will continue to emerge whose missions are centered first and foremost on climate, while existing organizations with related pre-existing missions will continue to embrace climate as "their" issue and, as public consciousness about climate rises, will be increasingly influential on that issue. But if we think of the existing constellation of environmental NGOs as an important piece of societal infrastructure, it is an infrastructure that was not really built for this purpose—and in which positions on climate can be constrained by organizational history and perceived mission.

At the same time, support for addressing climate can also come from other unrelated—and somewhat unexpected—sources. Two governmental organizations, in addition to the resource management organizations discussed in an earlier chapter, fall into this category. The U.S. Department of Defense recently released an important report, in response to a Congressional request, about the national security implications of climate change (Department of Defense 2015). Their conclusion was that climate change has the potential to further weaken politically unstable areas of the world, limiting the ability of some governments to address the needs of the populations they govern. The interests of environmentalists and those of the Defense Department have rarely coincided so neatly.

And the U.S. Centers for Disease Control and Prevention (CDC), through its Climate and Health Program (launched in 2009), tracks deaths and healthcare costs associated with climate and offers adaptation assistance to states and cities (Centers for Disease Control and Prevention 2015). They project climate change will result in increases in heat-related and weather-related deaths and injuries, changes in the ranges of disease-causing organisms and disease-vector organisms, problems with the safety and availability of food and water, even forced migrations. (For sobering details, see Global Change Research Program 2014, also from a U.S. government source). One would think more people would pay attention, since health is an issue of such broad concern. It is also tempting to speculate what will happen to these reports if a climate-skeptic conservative should ever gain control of the Executive Branch of U.S. government.

However, these conclusions are unlikely to reach the average information consumer unless they go looking for them, and even then both their search and their interpretation of what they find likely depends on their political ideology and their trust in particular kinds of sources. Most science journalism is no different from other journalism in tending to be event-driven rather than issue-driven. Thus, between big storms, absent major new findings, after large international meetings have become past history, and during times in which other crises that seem more immediate push climate off of the news agenda, we're not likely to hear very much about it. Climate, as a future disaster although clearly portended by present ones, becomes easy to ignore.

New Audiences: Active Information Seekers

Given the great wealth of information available via both traditional and Internet sources, citizens wanting to learn more about climate (or any other complex topic) must become wise information consumers who know what they are looking for and are motivated to seek it out. What turns people into active information seekers for any given issue, in particular issues involving risk? On this question, scholars already have a good beginning. The Risk Information Seeking and Processing Model (or RISP) proposed by Griffin et al. (1999) combines concepts from several earlier models of information processing, including the idea of "information sufficiency" or how much information people feel they actually need, "information capacity" or how well they feel they are able to comprehend the relevant facts (both adapted from the Heuristic-Systematic Processing Model of Eagly and Chaiken 1993), "channel beliefs" about available information channels such as the news media (adapted from Kosicki and McLeod 1990), emotional or affective responses, and "informational subjective norms" or what people think others expect them to know, a variation on a key element in the Theory of Planned Behavior (TPB) that addresses the expectations of others (Azjen and Fishbein 1980). Characteristics of the hazard in question and of the individual figure into this equation as well.

This is a complex puzzle that has been tested and extended in many contexts, too many to be fully explored here. For the present discussion, the primary importance is broadly conceptual: People seek information that they feel they need, or that they feel others expect them to know, and their seeking behavior depends on many factors, including their beliefs about particular channels of information (a concept related to the idea of "source credibility", which has a long history in mass communication research), as well as trust, a key variable in many risk communication studies. It is something of an irony that this basic model was developed at a much earlier stage in the development of the modern Internet, which ultimately led to radically expanded information choices from a user perspective. The relevance of the RISP model is even more apparent today than when it first appeared.

Of special interest for this book, the model has been applied to support for climate change mitigation policy, as alluded to in an earlier chapter (Yang et al. 2014). In that study, two newspaper-style stories—one discussing the idea of a government-imposed carbon tax (collective action) and one discussing the possibility that individuals could use less air conditioning (personal action)—were tested in an online experiment with student subjects. People who engaged in what is called "systematic"—that is, thoughtful—information processing (in addition to seeking itself, generally predicted as an outcome variable in the RISP model) were more likely to be supportive of mitigation measures. However, the authors found "disconcerting" the conclusion that those who already felt confident of their climate knowledge might only process the information "heuristically"—that is, superficially—since this might indicate over-confidence and a missed opportunity to learn more (Yang et al. 2014: p. 317).

In much of the communication research literature, including the RISP literature, so-called systematic or thoughtful information processing generally emerges as superior to heuristic or cue-based (superficial) processing, in terms of message impact. So if we imagine someone standing in a library or bookstore (or sitting at a computer screen, of course) scanning through book after book (or website after website), some will be only glanced at because the title is not appealing, the author doesn't seem to have the right demographic characteristics, or the seeker is simply not attracted to the topic—or even to the cover. Any of these could lead the seeker to either accept or discount the information on superficial grounds, rather than on the basis of careful (systematic) thought. Whatever information encourages more systematic processing—deeper thought, that is—is likely to be more influential in the long run.

While systematic processing is generally seen as a positive outcome, heuristic processing is not always a bad thing; in fact the Yang et al. study described above found no relationship between heuristic processing and support for policy changes, rather than a negative relationship. It is worth stressing that, especially in this era of information overload from all kinds of electronic media, we all very often need ways to focus, limit, and otherwise manage our information seeking. We are forced to manage some decisions by "satisficing" (Simon 1956)—that is, getting just enough information to get by on, in order to make a decision or take a position. Human beings are complex, and both heuristic processing and "satisficing" may be perfectly rational strategies under some circumstances. But which circumstances are those? Some people might need only a limited amount of science in order to move forward with action; others may feel a need to understand as much as possible first.

As these authors note, and as Chap. 3 of this book suggested, the communication research literature also suggests that information likely to arouse negative emotions, such as fear, should be accompanied by information on efficacy—how one can take steps against the threat. This lesson was first learned in the context of health communication. Without steps to be taken, fear can simply encourage people to "shut down" or reinforce their denial (whether the subject is climate change or tobacco smoking). For climate, efficacy information seems as though it should be vital—not only to lessen the potential impact of the kind of fearful response that might trigger denial, but also as a means to encourage people to take the next step: positive action. This won't always happen as expected; while factors such as environmental values and personal relevance matter, so do tendencies toward confirmation bias (seeking information that matches one's existing beliefs) and motivated reasoning (developing rationalizations from those beliefs). Further, some people may feel they have enough information when they *understand* climate, whereas taking action requires more—and a different kind of—information. And since combating climate change will take a broad global effort, efficacy information targeted at individuals may not make enough impact.

Both the RISP model and the Theory of Planned Behavior (TPB), which considers behavioral outcomes more generally, have the advantage that they incorporate a very important collective dimension: the expectations that others have of what we ought to know (or, for TPB, what we ought to do). Note that RISP—designed to predict information seeking—defines efficacy primarily with respect to informational factors, whereas TPB—designed to apply to a broader class of problems—defines a sense of efficacy largely as whether taking action is possible and will have impact. One TPB-based study of climate change messaging has shown that including efficacy information will increase hope across liberals, moderates, and conservatives, although these groups differed in their responses to various forms of efficacy, and that hope in turn increases political activism intent (Feldman and Hart 2016). Fear also seems to depress activism intent in some forms and among some groups. Such outcomes could be interpreted as suggesting that just as for health promotion messages, a sense of efficacy might prevent a feeling of hopelessness that could otherwise undermine an intent to act, whereas fear might do the opposite. More research is needed on all of these points.

Testing another form of activism in a very practical context, intent by homeowners to participate in a home energy conservation "upgrade" program, Priest et al. (2015) showed that participation intent responded not only to the prospect of financial savings through conservation but was also influenced by environmentalist values. Based in a general way on diffusion theory (Rogers 2003), as well as the TPB, the study looked at intention to adopt conservation measures such as improved insulation or more efficient appliances, steps that require homeowners to make a financial investment. While 95 % of respondents to this mail survey, conducted in the state of Nevada, said that saving money on utility bills was important or very important to them, only 4 % intended to take action along these lines in the near future. The best predictors of actual intention to act were an environmental orientation and a positive attitude toward the idea. Simply classifying people as early or late adopters, the basis of diffusion theory in its original form, is not enough; we also need to take into account values and attitudes.

The information environment has become far more complex, yet emotions, values, attitudes and expectations—certainly not just scientific facts—predict climate action. In the next two chapters we will address two interrelated questions: what both journalists and citizens actually need to know to make sense of scientific findings in a world in which traditional gatekeepers are much less important than they were only a few decades ago (Chap. 6) and why climate change seems to keep disappearing from the media—and the public—agenda, as well as how this might best be addressed (Chap. 7).

Note

1. The AAAS membership includes other professionals with science-related interests, including communication specialists, policy specialists, administrators, and many others, not just practicing scientists. It should therefore not be considered perfectly representative of scientists. And because AAAS is in part an advocacy organization, these results may well overstate (rather than understate) the proportion interested in public communication.

References

Adler, B. 2014. Why is Environmental Defense Fund Backing Lindsey Graham? *Grist*. http://grist.org/politics/why-is-environmental-defense-fund-backing-lindsey-graham/

Azjen, I., and M. Fishbein. 1980. *Understanding Attitudes and Predicting Social Behavior*. Prentice-Hall.

Barthel, M. 2015. *Newspaper Fact Sheet*. Pew Research Center. http://www.journalism.org/2015/04/29/newspapers-fact-sheet/

Centers for Disease Control and Prevention. 2015. Climate and Health. http://www.cdc.gov/climateandhealth/default.htm

Department of Defense. 2015. DoD Releases Report on Security Implications of Climate Change. http://www.defense.gov/News-Article-View/Article/612710

Eagly, A. H., and S. Chaiken. 1993. *The Psychology of Attitudes*. Harcourt Brace and Janovich.

Feldman, L., and P.S. Hart. 2016. Using Political Efficacy Messages to Increase Climate Activism: The Mediating Role of Emotions. *Science Communication* 38(1): 99–127.

Fogg-Rogers, L., J.L. Bay, H. Burgess, and S.C. Purdy. 2015. "Knowledge as Power": A Mixed-Methods Study Exploring Adult Audience Preferences for Engaging and Learning Formats Over 3 Years of a Health Science Festival. *Science Communication* 37(4): 419–451.

Gandy, O. 1982. *Beyond Agenda-Setting: Information Subsidies and Public Policy.* Ablex.

Gerken, J. 2012. "I Vote 4 Energy" Video Spoofs American Petroleum Institute Ad Campaign. *Huffington Post,* January 5. http://www.huffingtonpost.com/2012/01/05/i-vote-4-energy-video-spoof-api_n_1186400.html

Global Change Research Program. 2014. *National Climate Assessment.* http://nca2014.globalchange.gov/report

Griffin, R.J., S. Dunwoody, and K. Neuwirth. 1999. Proposed Model of the Relationship of Risk Information Seeking and Processing to the Development of Preventive Behaviors. *Environmental Research* 80(2): S230–S245.

Hamlett, P., M. Cobb, and D. Guston. 2008. National Citizens' Technology Forum: Nanotechnologies and Human Enhancement. *Report No. R08-0003.* Arizona State University, Center for Nanotechnology and Society. https://cns.asu.edu/sites/default/files/library_files/lib_hamlettcobb.pdf

James, B. 2015. Forget Cable Cord-Cutting: 83% of American Households Still Pay for TV. *International Business Times,* September 15. http://www.ibtimes.com/forget-cable-cord-cutting-83-percent-american-households-still-pay-tv-2081570

Kosicki, G. M., and J. M. McLeod. 1990. Learning from Political News: Effects of Media Images and Information-Processing Strategies. In *Mass Communication and Political Information Processing,* ed. S. Kraus, 69–73. Erlbaum.

Lee, M., and M.S. VanDyke. 2015. Set It and Forget It: The One-Way Use of Social Media by Government Agencies Communicating Science. *Science Communication* 37(4): 533–541.

Lewenstein, B. 1995. From Fax to Facts: Communication in the Cold Fusion Saga. *Social Studies of Science* 25(3): 403–436.

McWright, A.M., and R.E. Dunlap. 2011. The Politicization of Climate Change and Polarization in the American Public's Views of Global Warming, 2001–2010. *Sociological Quarterly* 52: 155–194.

Meyer, M. 2010. The Rise of the Knowledge Broker. *Science Communication* 32(1): 118–127.

National Center for Education Statistics. 2015. Digest of Education Statistics: 2013. *Report NCES 2015-0011.* United States Department of Education. http://nces.ed.gov/programs/digest/d13/index.asp

Newport, F. 2012. Americans, Including Catholics, Say Birth Control is Morally OK. http://www.gallup.com/poll/154799/americans-including-catholics-say-birth-control-morally.aspx

Orenstein, D. 2009. Futurity, an Online Outlet for Research News, is Launched by Stanford and Other Leading Research Universities. https://biox.stanford.edu/highlight/futurity-online-outlet-research-news-launched-stanford-and-other-leading-universities

Priest, S., T. Greenhalgh, H.R. Neill, and G.R. Young. 2015. Rethinking Diffusion Theory in an Applied Context: Role of Environmental Values in Adoption of

Home Energy Conservation. *Applied Environmental Education and Communication* 14(4). 213–222.

Rainie, L., C. Funk, M. Anderson, and D. Page. 2015. How Scientists Engage the Public. http://www.pewinternet.org/files/2015/02/PI_PublicEngagementby Scientists_021515.pdf

Rogers, E. M. 2003. *The Diffusion of Innovations.* 5th ed. Free Press.

Saba, J. 2009. *Editor & Publisher.* Specifics on Newspapers from 'State of News Media' Report. http://www.editorandpublisher.com/news/specifics-on-newspapers-from-state-of-news-media-report-2/

Scheufele, D. 2011. Modern Citizenship or Policy Dead End? Evaluating the Need for Public Participating in Science Policy Making, and Why Public Meetings May Not be the Answer. *Research Paper Series No. R-34.* Joan Shorenstein Center on the Press, Politics and Public Policy. http://shorensteincenter.org/wp-content/uploads/2012/03/r34_scheufele.pdf

Sierra Club. n.d. Wind Siting Policy. http://www.sierraclub.org/policy/energy/wind-siting-advisory

Simon, H. 1956. Rational Choice and the Structure of the Environment. *Psychological Review* 63(2): 129–138.

Song, L., and K. Bagley. 2015. EDF Sparks Mistrust, and Admiration, with Its Methane Research. *Inside Climate News.* http://insideclimatenews.org/news/07042015/edf-sparks-mistrust-and-admiration-its-methane-leaks-research-natural-gas-fracking-climate-change

Weingart, P. 2001. Die Stunde der Wahrheit? Zum Verhältnis der Wissenschaft zu Politik, Wirtschaft und Medien in der Wissensgesellschaft *[A Moment of Truth? The question of Science's relation to Politics, Economy and Media in a Knowledge Society].* Velbrück.

Wynne, B. 1989. Sheepfarming After Chernobyl: A Case Study in Communicating Scientific Information. *Environment* 31(2): 10–39.

Yang, Z.J., L.N. Rickard, T.M. Harrison, and M. Seo. 2014. Appling the Risk Information Seeking and Processing Model to Examine Support for Climate Change Mitigation Policy. *Science Communication* 36(3): 296–324.

CHAPTER 6

Critical Science Literacy: Making Sense of Science

Portions of this chapter are adapted from material that was originally published by the author in the June 2013 issue of the Bulletin of Science, Technology & Society as Critical Science Literacy: What Citizens and Journalists Need to Know to Make Sense of Science. 33 (5–6): 138–145.

What does it mean to propose that systematic processing of scientific claims and related issues is preferable to heuristic processing? Once people search for and find information about climate change, what should they do with it? This chapter takes a look at the complexity of the background knowledge and assumptions that influence the interpretation of claims portrayed as scientific, concluding that it would be helpful to think of science literacy in a different way. This observation may not lead directly to a "fix" for climate denial, but it contributes to our understanding of the challenges that both science communication and science education need to address, going forward. Democratic discussion of climate solutions will ultimately need to rest on some common ground. We suggest here that part of that common ground will need to be an ability to consider scientific claims critically—that is, thoughtfully—and with some level of understanding of where those claims come from—that is, how science actually works. The shift from "deficit" to "dialogue" and the free-for-all Internet information environment we inhabit both ask much more of individual citizens who (whatever various expert communities might prefer) will ultimately choose for themselves which claims to believe.

First, we raise some issues here about the way scholars have looked at this question in past communication research. We know that people's values and even ideologies come into play in forming opinions about climate. Political ideologies in particular have proven a substantial barrier to accepting the science about climate. Yet ideology is not always irrelevant to climate-related decisions. People who fully accept the scientific reality of climate change may not "buy into" some proposed solutions (e.g., nuclear power in energy policy or a carbon cap-and-trade economic policy), and a reasonable person might indeed base their policy choices on values and ideology rather than strictly on the science. In other words, relying on values and ideology is not necessarily bad or wrong. Further, it is also inevitable. Using one's personal values to decide between different interpretations of available scientific evidence may not be a good idea, but using one's values to *decide what to do* on the basis of that evidence seems to be the right path—and it is a path that should involve active and engaged thought and discussion, rather than simply political polarization.

Scholars looking at how people process information generally assume that systematic processing is preferable because it is more likely to lead to enduring learning and even opinion shifts in a direction considered desirable, while heuristic or cue-based processing is based on superficial cues—sometimes resulting in the kind of knee-jerk reaction that simply overrides the facts, arguably the worst kind of heuristic processing. In actuality, the line between these two kinds of cognitive processing is blurry; they might best be thought of as a continuum rather than a dichotomy (Seethaler 2016). For science, this line is particularly difficult to draw. Since scientific data never really "speak for themselves", a certain amount of heuristic processing is necessary for non-specialists to appropriately process information from science. Citizens need to decide who should be trusted as a scientist source or other spokesperson, a judgment made more difficult by the fact that in science, part of keeping an open mind means that the theory that turns out to be correct may come from an unlikely source. Figuring out what science (and which scientists) to trust without relying too heavily on simplistic heuristics is not so easy. Cues like institutional affiliations, degrees earned, and consistency with what other scientists are saying do, in fact, matter to this task.

Up to this point, much of the discussion in this book has tried to highlight what is collective or social about human communication—whether about climate or more generally. Human communication as we understand it cannot take place on a desert island inhabited by a population of one. Humans are inherently social creatures that do not usually form enduring opinions in a vacuum—or, for that matter, in a laboratory. The human values and beliefs underlying our opinions are generally embraced collectively, although there is plenty of room for variation within and across groups in our pluralistic society. The move from "deficit" to "dialogue" in science communication represents, in part, a recognition that thoughtful discussion is often a valuable complement to individual reflection. Clearly, both play a role. Science itself is a highly social activity, as this chapter will elaborate: It represents another case where the human whole is greater than the sum of the individual parts.

As powerful as the climate of public opinion (real or perceived) seems to be, moving toward meaningful action on climate will also require an ability to grasp and accept some essential scientific facts. And as attractive as the idea of more democratic discussion of science might be, there is a "hard limit" on the extent to which science can be democratized: Scientific ideas can certainly be contested, and they can be wrong, but their truth cannot be determined by majority vote—or even by extended discussion, necessarily.

An assumption of this book, sometimes implicit, has been that humans are inherently rational (that is, they are universally capable of deeply rational thought), and that when people appear to be irrational, they are probably accessing different information and processing it differently, often through the lens of a different set of values and beliefs. Nevertheless, the assumption that people are at root rational seems necessary; otherwise, democracy itself is a tenuous proposition. It follows as a sort of corollary that in a pluralistic society especially, sometimes some people's beliefs need to be worked around, since they are unlikely to change. We hope that this is true for only a small proportion of "deniers", what spiral of silence theory thinks of as the "hard core", and that most others will eventually get on board with the mainstream view, but it is difficult to be certain how this will turn out. In our democracy, we value both freedom of belief and freedom of speech. Suppressing dissent would violate those core values, so we often have to move ahead incrementally and without universal consensus. Yet for climate, time is increasingly getting short.

While knowledge of what climate science has to say may be in itself insufficient to persuade people of the reality of climate change, perhaps more widespread knowledge of "how science works" (in terms of its social organization) would help. With the "gatekeeper class" of science journalists shrinking and anti-regulation ideologues cynically attempting to minimize market interference by promoting climate "skepticism", now is a good time to reconsider the nature of science literacy. This is not a retreat to "deficit model" thinking but involves the recognition that scientific facts, narrowly defined, are not the only legitimate basis of our attitudes and opinions on science-related policy issues, and that how people think about the enterprise of science is indeed relevant. Are scientists trustworthy—and which ones? Why are scientists unable to predict the weather but feel they are able to predict the climate? Are they actually trying to fool us? Uncertainties and associated probabilities in science are inevitable. What do most people make of those? When people seek and then find climate information, what do we expect them to do with it?

Heuristic processing of information about scientific findings (while not always something to be avoided altogether) is consistent with how science is often taught, that is, as a collection of facts to be memorized but not particularly analyzed, despite many attempts to reform science education. Even so, as noted earlier, heuristic processing of science-related information is actually essential for all of us, at least some of the time. We are mostly not scientists, and even those of us who have PhD degrees have expertise that is—virtually by definition—limited in scope. Scientist or not, we cannot always verify science-based claims directly (unless we are a true scientific expert in the field in question). Rather, we have to decide which sources and which interpretations to trust. Odd it is, then, that raw data on climate is so commonly held up to scrutiny in the popular press.

Science literacy is often defined (or at least measured) as awareness of a specific collection of important scientific facts—the right answers to particular questions. This is extremely short-sighted (if, for some purposes, administratively very useful and convenient). Even to define science literacy as a skill set is not entirely sufficient; people need the skills to evaluate scientific claims, but these are not necessarily the same skills that scientists use to produce those claims, and knowledge of scientific method is just one of the relevant factors. Part of the background knowledge needed to identify valid scientific claims is familiarity with a number of important ideas, such as recognition that some uncertainty always surrounds specific scientific claims, understanding the nature of scientific specialization and expertise, familiarity with the range of available methodological approaches—from observation and classification to experiments and model-building—and awareness that the conduct of science is itself a social process. None of these is part of the canonical knowledge tested when we test science literacy, but perhaps some of it should be. This is true even though some of the relevant elements—the reputations of particular scientific institutions, for example—may seem more heuristic than systematic.

Debate about climate-and-energy policy, both that addressing mitigation and that addressing adaptation, is essential. The future choices we will make as a society will involve political values, environmental values, and even aesthetic values (e.g., with respect to wind farms). They will also involve our ideas about fairness, justice, and human welfare, including our valuation of health and security. So we should not be tempted to conclude that bringing values or ideologies into the debate is inherently problematic. Using values and beliefs to determine scientific veracity is usually a problem,[1] but value-based thinking is to be expected in science-based policy making. Recognizing the problems stemming from climate change and developing solutions is a very value-laden proposition, and as a society we may therefore never agree on every step. We should find ways to encourage people to recognize the important role of values in these decisions in a way that encourages appropriate values-based thinking, something that it would be good to reflect in both news accounts and strategic messages about climate. This does not, of course, mean that scientific evidence can be interpreted any way one likes. But we should aim for a future in which the debate over choices is centered on what to do, not whether climate change exists. This might be facilitated by an agenda shift in public communication efforts. And as in all debates, informed participants are an essential component.

THE SOCIAL SIDE OF SCIENCE—AND WHY IT MATTERS

Science is social in many important ways, despite the "nerdy" reputation of scientists as people with limited sociability. To understand how science works, we need to think about the way that collaborations and mentorship work, about the structure and function of tenure and peer review systems, about the deliberations of funding bodies, and about the roles of academic societies and academic meetings, as well as universities, funding agencies, research journals, and other important research institutions, in fostering and overseeing our collective search for scientific truth. These are all social arrangements, managed (imperfectly, to be sure) under the auspices of important social institutions. This same kind of awareness of the social side of science is needed by those who write about science (and scientific claims) for public audiences, such as science journalists—and what they write has the potential to foster further growth of science literacy among their audiences. The term "critical science literacy" is proposed here to refer to this basic understanding of the social nature of science—including an appreciation of the diversity of methodological approaches within the scientific community and the inevitable role of uncertainty.

As a thought experiment, it is useful to contemplate what scientists themselves and other scientifically literate people do when evaluating a novel scientific claim, especially one outside their own expertise. They think about the credentials of the researchers, where the work was presented or published, and at what university (or government or industrial laboratory) the research was done. They likely also consider whether the conclusions are consistent with present paradigms: A scientific study that claimed to establish the existence of ghosts would be greeted skeptically, as would one that claimed to entirely overturn the theory of evolution. A scientifically literate person might well ask him or herself questions about funding sources and other possible biases, as well as whether the methods used seem—on the surface of things at least—to be "scientific". If they are even a little bit sophisticated about both science and journalism, and they are looking at a news story, they will look carefully to see if multiple expert sources have been consulted—or if there is other evidence of consensus among relevant expert communities.

Even scientifically literate people are not immune to being taken in by pseudo-scientific claims. Heuristic processing is not a "yes or no" proposition; there are many layers and levels, and it is also not categorically a bad thing. Again, we all engage in it to some degree; it is needed as a kind of triage to manage today's information overloads. But a scientifically literate person in the sense being described here would not be very much persuaded by an article written in an unknown journal by an uncredentialed researcher whose lab was in his basement and who purported to prove that the earth is, after all, quite flat—or at least we hope they would not be persuaded. A scientifically literate person should have some degree of immunity to believing bad science, in other words.

All of these cues (some of which can be thought of as heuristic) establish a basis for trusting or not trusting scientific results (or their messenger). Trust is a characteristic that is also, albeit erroneously, sometimes conceived of as a heuristic, purely emotional response, rather than a thoughtful, "rational", or systematic one. But trust is not just a thoughtless emotion. It also derives from experience with similar messages and messengers in similar contexts and—for science—from awareness of the nature of the scientific enterprise. Trust is often a perfectly rational consideration, in other words, not necessarily either a particularly superficial or a purely emotional one. In cases of science being reported in news media, these judgments are often made without a detailed evaluation of methods or hypotheses. That kind of detailed evaluation is largely limited to cases where far more details are available (as in a journal article or a lengthy academic presentation), a common complaint from scientists against news reports and their (necessary) brevity. For non-scientist audiences, however, an extreme level of detail may be simply a distraction. We have little choice but to trust both the scientist and the journalist bringing us a report about science, alongside attention to cues such as credentials and affiliations.

Even the systematic evaluation of the type that the peer review process for journal articles is supposed to offer is far from infallible. Experts cannot always detect misrepresentations by reading a journal article or other research report. They can only detect logical error, such as statistical fallacies; evaluate the reasonableness of the methods as reported; spot errors of omission, such as failing to consider directly relevant work published by others; and (in most cases, we hope) recognize outright quackery. In the end, they also have to trust the scientist as well to have done what it is that they claim they did. Results must then be verified through replication and often through discussion at scientific meetings as well. Scientific truth is ultimately distilled through collective processes. This is an imperfect process, especially in the short term—but it is arguably the very best one that civilization so far has had to offer with respect to understanding our physical environment.

In evaluating new scientific ideas as scientifically literate citizens, even as scientists, we also rely heavily on what the late sociologist Robert Merton enduringly described as the "ethos" of science introduced in Chap. 5: We generally assume that the conduct of the research was carried out by an ethical person engaged in a disinterested search for truth, in other words. If scientists were routinely engaged in making up data in order to fool other people, the entire system would quickly collapse; while this does happen, it is definitely not a routine occurrence. Of course, scientists can also be wrong, something that happens quite regularly. Knowing this is also a part of critical science literacy. Every scientific assertion remains tentative, pending new observations and new interpretations of existing data.

In reality, it is not conceivable that even scientists will evaluate most scientific claims they encounter—especially those outside their own field—by actually examining the data or questioning the procedural details beyond a fairly superficial look. The use of trust, with respect to both the individual scientists involved and the social institutions such as universities, journals, and scientific societies that are also involved, is absolutely essential. It is not a mistake or a lazy alternative; it is not an incidental characteristic of the way that science is evaluated. Rather, it is the only reasonable way that reasonable people (including scientists) can make sense of science on a day-to-day basis. If people have no rational basis for deciding which people making scientific claims are trustworthy and which are not, then climate change denial becomes a bit more understandable. It is not justifiable, it is counter to the collective scientific consensus, and it is wrong. But at least it can be understood.

With an economy and society dependent in so many ways on science and technology, some level of public science literacy seems a prerequisite of sustained success, both for individuals and for the collective. This goes well beyond current controversies and certainly includes success in terms of creating and sustaining jobs and businesses and managing personal careers, but it also includes success in terms of making wise choices, recognizing which claims might be suspect, knowing which voices represent bona fide scientific expertise and even which other voices are also relevant. The "public engagement" movement in science communication has long embraced the idea that it is not just scientists whose expert knowledge deserves respect, but rather that many forms of "lay" expertise often matter as well (see Wynne 1989).

The idea that scientific truth represents (in the ideal) an active social consensus among experts derived after extended evaluation and disinterested testing is central to understanding how science actually works. However, this is an element that is not necessarily widely understood. The lone scientist crying "Eureka!" over a test tube is part of the popular image of science, but not a group of scientists looking into a computer screen together at a fuzzy collection of dots—or arguing about just how those dots should be interpreted or how precisely the data underlying them has been measured.

So-called "normal science"—everyday scientific work—progresses incrementally (if sometimes in zigzag fashion) and generally as a result of the conduct of careful research by ethical people with appropriate expertise. Again, of course this is an idealized version of a less clear-cut reality, but it is a vision that matters. If scientists stretch an interpretive point to support a personal pet theory or please a funding source or attract more media attention, it is hoped—and it is often the case—that the scientific enterprise will generally self-correct as future research unfolds.[2] Rarely, as sociologist of science Thomas Kuhn (1970) showed, an underlying paradigmatic belief will be firmly upset and major progress may result all in one quick leap. But this is certainly the exception rather than the rule. Sometimes the day-to-day progress of science is not particularly exciting, truth be told. New science can also can take quite a while to permeate scientific thinking and even longer to engender societal awareness; according to historical accounts, climate change was originally discovered and connected to human-caused CO_2 emissions well over one hundred years ago, in 1896 (see American Institute of Physics 2015).

At least in the news, however, the story of scientific progress seems continuously punctuated by major breakthroughs. Part of the blame rests on the single-study press releases commonly distributed by universities—and occasionally by scientists themselves, or those claiming to be scientists. Prominent examples of scientific misrepresentation that come immediately to mind include the "cold fusion" controversy discussed in the previous chapter, apparently based on premature results; the apparently fraudulent Hwang human stem cell cloning claims published in *Science* in 2004 and 2005, resulting in both papers being retracted (for details, see *Science*/American Association for the Advancement of Science 2016); and the New Madrid earthquake prediction made by a man named Iben Browning in 1990 (Spence et al. 1993), which sent droves of reporters to a small town in Missouri to cover an anticipated quake (one that never materialized) predicted by someone who has been variously described as a biophysicist, a climatologist, and a zoologist—but who was not a credentialed geologist. (And even credentialed geologists cannot yet make such precise predictions about earthquakes, of course.) More recently, the original scientific paper asserting that autism might result from vaccination was withdrawn (Harris 2010). Yet some distrustful parents still cling to a theory made to seem more credible by "false balance" reporting" (Clarke 2008).

Perhaps news consumers should not be blamed for being suspicious of new claims, wherever they arise. Institutional legitimacy, while important, is not a sure guide. Pons and Fleischmann spoke to the world about cold fusion from a press conference at a recognized U.S. university; Woosuk Hwang's work on stem cells and human cloning was introduced with great fanfare at one of the most prominent scientific society meetings in the world (AAAS); and Andrew Wakefield's autism–vaccine paper had been published in *The Lancet*, a prestigious British medical journal. But the bad news about climate change is hardly new. Some reasons for ongoing resistance to modern climate science have been offered in previous chapters, from anti-regulation ideology to blanket denial based on fear to deliberately planted seeds of doubt. It seems likely all three explanations are correct. Our tendency to treat science as a collection of immutable facts to be learned by rote should be added to the list. Just because the science of climate change is still emerging, still being refined, and is still marked by uncertainties (e.g., surrounding rapidity and local effects) does not mean it is unscientific.

Whose Fault Is All This Confusion?

In many of these and similar cases of manifest misrepresentation and public confusion surrounding scientific claims, the news media are regularly blamed—rightly or wrongly—for journalistic errors and their apparent consequences in terms of public perception and misperception. It is all too true that conflict, fraud, and radical new results all make attractive topics for journalists. It is also true that some journalists make stories out of university press releases without too much concern for verification or providing needed context, especially journalists working for smaller organizations and in resource-stressed situations. However, sometimes the scientific truth of things is simply difficult to discern, whether for scientists, journalists, or their audiences. Suffice it to say that journalists also need a good dose of critical science literacy, including what is sometimes called "healthy skepticism". It is not enough to introduce new claims by a qualifying phrase such as "scientists reported"; the very appearance in the news of unchallenged assertions lends them a legitimacy that should be earned. But once earned, and the reality of climate change has certainly earned its legitimacy, it is irresponsible for journalists to pretend otherwise absent a shift in the scientific consensus.

Today, in the Age of the Internet, everyone can be a journalist, creating further opportunities for the dissemination of diverse claims and counter-claims about scientific truth, as about all truth. This may ultimately be very good for democracy, even very good for public education longer term, but it creates an environment in which it can be more challenging than ever to decide whose truth ought to be believed. The Internet has created, in other words, a landscape that can be difficult to navigate. While one might hope that a scientifically literate citizen will not be easily misled, this hope may sometimes be misplaced.

The contemporary media environment is one where we all have access, through both "legacy" and "social" media, to a steady stream of entertainment, promotion, information, news and advertising, one in which these elements can no longer be readily distinguished from one another. Fortunately or not, current generations seem quite comfortable jumping from one mode of presentation to another, and they navigate the web world with great apparent skill. Science is everywhere: on Twitter, on Facebook, on YouTube. These are important new opportunities for science communication. But what science literacy might mean in this context—a world where every factual question, from the trivial to the profound, seems answerable with a simple search—is overdue for further thought.

Only a fine line divides false (or simply unsubstantiated) claims from claims that are actively fraudulent. On the surface of things, it seems likely that Utah scientists believed that they had discovered something fundamentally new, which they described as "cold fusion". Given subsequent events, including the revelation that Hwang had gotten some egg donations from his female lab assistants and paid for others, whereas he had claimed the donors were all volunteers, it seems plausible that his reports of human stem cell cloning success were deliberately manufactured, rather than representing some sort of laboratory mistake. Iben Browning's motivation is less clear; perhaps he merely sought notoriety, or perhaps he believed firmly in his own earthquake predictions and was simply deluded about their validity and reliability. Sorting out which is which is not easy, even years after the fact. Note that such judgments involve (in each and every case) imputed character and not just the scientific evidence. "Can we trust the scientists?" is actually a perfectly reasonable and relevant question, doubly so for those without trusted personal connections inside the scientific community.

The oft-implied expectation that science journalists (let alone other non-scientist citizens) should have sorted out these claims better in the first place, especially since the scientific community had not sorted them out yet either, is something of a tall order. Sure, it is certainly possible to imagine journalists behaving in a more responsible way. Perhaps it is reasonable to expect that journalists could at least check into people's formal degrees, as a necessary if not always sufficient indicator of possession of relevant expertise, but that does not in itself either prove or disprove whether their conclusions in a new area of science are true or false. In principle, valid new scientific discoveries could come from anyone. Again, the system assumes that discussion, debate and replication should sort this out. But a full century into studying climate change and its human origins, some facts should be accepted as legitimate until proved otherwise, rather than uncertain until more positively proven.

Climate change is an especially complex case, in part because it is simply difficult to accept. For some people, it likely upsets their worldview in a variety of disagreeable ways, just as much as upsetting the theory of evolution would distress most biologists. In addition, substantial funding is being directed toward encouraging "skeptical" (that is, "denier") voices in an era of unprecedented information flow. The application of the journalistic norm of "balance", borrowed from political reporting where there is generally a "left" and a "right" position on most issues, to climate change—as for other scientific controversies, real or perceived—provided a convenient opening for both ideologically motivated "deniers" and others who believe delay in addressing climate will be in their own self-interest. Today, via the Internet, there are many more opportunities to post contrarian ideas. Some of these contrarian voices appear to have relevant scientific credentials; investigation is often required to confirm this, rather than a superficial look, and while most journalists are certainly not unintelligent, they work in an increasingly stressed environment where time may not permit even that elementary sort of scrutiny. Some contrarians may be legitimate objects of public attention, even if they are not scientists and do not speak for science—another very fine line. And today's news consumers get much of their information outside of traditional media—arguably a good thing, but one calling on those consumers to exercise critical judgment in new ways.

Not all attitudes toward science-related issues rest primarily on understanding scientific facts; objections to stem cell research, for example—whether or not we agree with them—are not based on misunderstanding of science and are unlikely to be resolved by more science. This may also be the case for some anti-vaccination voices, although in that case (and many others, including climate) non-scientific arguments seem to be deeply entangled with misunderstanding of the underlying science, sometimes motivated misunderstanding. Not to cover dissent smacks of censorship, though, and citizens in a democracy should consider as many points of view as possible, even wrong ones; they should know about things like refusal to vaccinate, just as they should know about other conflicts in society, and ignoring dissent seems more dangerous to our system than at least reporting it and letting people make up their own minds. Yet to reasonably interpret dissent within science, or dissent about science, requires the skills of critical science literacy.

When the media cover a controversial development in science or its application, we expect them to include comments from those who are less than enthusiastic about it, even if those comments do not come from credentialed scientists. For any given issue, religious, political, ethical, patient advocate, and/or environmentalist voices may also be relevant. It is reasonable to include these and other "lay" voices in journalistic accounts (Secko et al. 2013) and to report the existence of controversy. When citizens protest a new nuclear power plant being built "in their backyards" (as the saying goes), we expect journalists to take note, albeit without necessarily taking sides. These debates are often not about good science but about wise and inclusive public policy, and citizens do have choices to make. Even for an established scientific truth like climate change, journalists would be rightly criticized if they completely ignored the existence of dissent. This is legitimately newsworthy. Yet as we all watch the glaciers melt, a phenomenon made vastly more visible by this selfsame expanded media environment, we have to ask questions about why so many ordinary people find it so difficult to sort out this particular truth, as well as what is the responsible way to discuss both dissent and the inevitable persistence of scientific uncertainty.

Contrarian views from outside science shouldn't routinely be represented as scientific—and yet controversy does exist within science, not just between science and society. Who is to decide what "counts" as "proper" science, especially when the science in question is still emerging? It is an underlying principle of scientific thinking that even some contrarian claims could potentially turn out to be true and that today's contrarian claim can (in principle) turn out to be just as correct as the historical recognition that the earth is spherical rather than flat. Open-mindedness and the idea that truth is subject to revision in the light of evidence are—quite legitimately—part of scientific tradition, and their very normalcy is also a necessary part of science education. Even so, we hope that personal and policy decisions that have to made at any given point in time can be made in an "informed" way—that is, taking the best available expert opinion into account, even if few truths can be deemed certain or eternal. We had best not ignore the underlying hard reality of something like the anthropogenic causes of climate change, anymore than we can afford to ignore the underlying reality of the relationship between smoking and lung cancer—or between vaccination and susceptibility to disease.

The idea of critical science literacy extends beyond trusting the right sources, however. Understanding "how science works" in practice means going beyond journal article, textbook and classroom idealizations of the processes that actually go on in laboratories and other research settings. As Latour and Woolgar's well-known study of "laboratory life" first established (1986), the reality rarely matches the ideal. As anyone who has actually done science or observed it closely knows, it can be a messy process. Data do not turn out to look like we expect, experimental equipment and procedures are imperfect, bad weather interferes with field work, and human beings make mistakes that cannot always be corrected. In the end, it is those same imperfect human beings who must decide what the results of an investigation actually mean.

This isn't always as clear cut as post hoc discussions may imply. An argument can be made that it is only by actually conducting science, side-by-side with scientists, that non-scientists begin to understand this messiness—and understand the value of science, despite its inherent messiness. This is a strong argument for "citizen science" projects in which non-scientist volunteers participate in data-gathering and analysis. It may also underlie (for some) the thinking behind high school and undergraduate laboratory classes, but in practice those are often a weak and artificial substitute for actually experiencing science-in-the-making. This is not to say, by any means, that the practice of science is "sloppy". It just doesn't necessarily follow textbook idealizations of scientific process, and journal articles are generally written in a style that extracts such a lean essence from the results that their imperfect derivation is readily forgotten—even within science. These imperfections are supposed to be "outed" through ensuing work. This is not a perfect system, but it is beyond doubt the best one we have. Alas, this process doesn't often work particularly quickly.

Redefining Science Literacy

Many other practical and theoretical criticisms can be raised about our existing approach to understanding science literacy. Almost any of these criticisms could take this discussion down one of several different paths deep into debates in the philosophy of science and the nature of human perception. What should "count" as knowledge or truth? Which facts matter, and who gets to decide what they are? Which aspects of our perception of our physical reality have to do with the "hard facts" of that reality, versus what we've been taught or tacitly assume? But the present discussion is less ambitious and concerned with a more practical question: What knowledge about science is of most central value to citizens in a contemporary democracy, in which many personal and policy decisions have some relationship to science or technology and most of the facts, observations and conclusions of science (and also of pseudo-science) are available so readily to us on our computer screens? How should citizens evaluate what they are told about climate change? What do people need to know in order to sort out which truths should be relied on in any given moment? This is clearly distinct from familiarity with the kind of specific, usually long-established scientific facts learned in middle school, most of which eventually go out of date.

The measurement of science literacy (in the old sense of "factual literacy") has been carried out by the National Science Foundation in the United States and similar agencies elsewhere for many decades. Over the years the use of true/false tests based on factual knowledge has persisted, even though understanding of the limitations of this approach has become more widespread. One narrow but valuable criticism is that some of these questions can muddle up knowledge with beliefs, e.g., in the standard question on evolution (Rughinis 2011). If the purpose of these tests is to establish whether science education is working, a measure that reflects religious beliefs might not be the best choice methodologically. And the wording of such items—as for all survey research—can be misleading.[3]

But the most obvious criticism is arguably more radical: the idea that no list of scientific facts quite captures what we hope citizens will bring to the disentanglement of situations like climate and the others described above. Citizens need skills that will serve them well when the facts are not yet clear, in other words. However, mastery of facts seems relatively easy to measure and it is administratively useful, so we tend to stick with it. And there is some cause for concern that we would not be able to track important progress without these measures. According to the 2014 National Science Foundation report on indicators of public attitudes and understanding, less than half of American believe that they understand "what scientists and engineers do at work" (National Science Foundation 2014). The average number of correct responses on a traditional test of science literacy was 5.8 out of 9, similar to previous statistics on this point, but fewer Americans than in past years (less than half) rejected astrology as "not at all scientific". Change in such indicators seems worth knowing about.

Measurement issues aside, many people who become concerned with science literacy in the traditional sense might assume—whether explicitly or implicitly—that higher levels of knowledge (understood as being in possession of correct answers to questions about accepted scientific facts, the same sort of criterion we ordinarily use to measure progress among college students and award them their grades) will translate to more support for science generally or more positive attitudes toward particular scientific or technological programs or innovations. Actually there is little or no proof of this "deficit model" assumption. Of course, just as Sturgis and Allum (2004) argued, knowledge does matter. It is just not the only thing that matters, and often it is not the main issue when science and society appear to be in conflict.

Today, as this book has discussed, many of those concerned with improving science communication have decided the goal should be to further "engage the public" in science, rather than trying to raise levels of factual knowledge among out-of-school adults by the seemingly more direct route of one-way information dissemination through lectures or documentaries. Public engagement in science and public discussion of the issues it raises have been promoted through science cafés, consensus conferences, public consultations, interactive museum events, and other similar forums. This kind of activity has been the subject of numerous programmatic experiments, a few of which have been discussed in previous chapters. The citizen science movement, which has followed a slightly different trajectory, promotes public engagement in the actual conduct of science—often in fieldwork settings. For example, a citizen network helps Cornell University ornithologists conduct annual bird counts (Bhattacharjee 2005). The U.S. National Oceanic and Atmospheric Administration (www.noaa.gov) provides grants for a variety of citizen science projects in marine contexts, as well as sponsoring citizen observations of local weather and climate through its Cooperative Observer Program and its Citizen Weather Observing Program. Such experiences may also do more to raise critical science literacy than any other form of education, and they do it entirely outside of classrooms. It is not clear whether proponents of citizen science approaches always see the purpose of their work in quite this way, but the trend toward "engagement" is valuable for multiple reasons—and few other approaches to this idea offer the same kind of direct participation in scientific research.

Those looking at "engagement" approaches as a sure path to convergence between expert and popular views are very likely to be disappointed, but they are steps in a good direction even so. Participation will never include everyone, but perhaps those who do participate can act as opinion leaders for science to the rest of us. Non-scientist citizens in our science-and-technology oriented society should learn more about science and explore how best to govern it. However, the sometimes implicit expectation that better collective decision-making will result from more citizen engagement, which rests on the quite reasonable assumption that "ordinary"" citizens have the intelligence and wisdom to reach good conclusions, may or may not be realized—especially in the short term. The hoped-for result that more "engagement" will result in wiser policy decisions remains elusive. Participants in such programs are generally already attracted to science; "engagement" projects won't easily engage the unengaged, in other words. There is also no obvious way to feed the outcomes of citizen discussions back into actual policy making—not surprising, given that the input from science itself into policy making is quite weak to begin with.

This latter is a political problem, more a problem in governance rather than a problem in science literacy. However, problematically, "engaging the public" is sometimes viewed as a strategy for science to gain public support as much as an attempt to improve democracy with respect to science-related choices. These two goals (encouraging engagement and generating support) are not the same; they may not even be compatible (Priest 2013). Further, given limitations of time and interest, in the end most people will still need to draw conclusions based on other forms of public communication, primarily mediated communication. The "engagement" approach may be our best current hope of raising the general awareness of how science actually works, but it seems unlikely ever to reach the lion's share of the public directly. We also need a new and more "critical" (in the sense discussed here) form of science journalism, one that concentrates more on science as a social phenomenon. This does not mean showing that science is wrong or somehow in need of reform, but only that it is human. Its strength lies in the strength of the social processes that sustain it: consensus, discussion, dissemination, evaluation, review, replication, comment, and even dissent. Too little of this is typically visible when scientific findings are reported.

What non-scientists generally want and need to know are which facts, observations and conclusions are most valid (and most reliable and most relevant) for supporting our individual and collective decisions, and the way that science-savvy audiences actually determine that—in practice—is closely related to understanding of how science actually works as a set of social institutions and practices that are governed by a set of normative and procedural assumptions. Science-related policy decisions are not very likely to arise for society with regard to well-established or "settled" science; they much more commonly concern new science and emerging technology to which our governmental and healthcare systems need to react. Under such circumstances, what should "count" as valid, reliable, and relevant truth is an especially slippery question. We cannot always answer that question in "real time". Meanwhile, climate change is progressing, and we must do our best to catch up with it. We simply cannot wait for every "denier" to be converted. A functional democratic system should allow for progressive action despite incomplete societal consensus.

To really understand the social enterprise of science, people need to know something about the sociology of science, as well as something about the philosophy of science. This is essential to navigating a world full of competing truth claims about science. While this may seem like a lot to expect, much of this knowledge is largely taken-for-granted among people who (for whatever original reason) became familiar with how science actually works at some point in their lives, whether this familiarity was the result of a career choice, an educational experience, or (say) a relationship with a scientist friend or family member. The flip side is that those possessed of some basic level of critical science literacy need to remember that others may not have this familiarity—discussions of science, both within and outside of the media, should not assume expert-level knowledge of scientific process or how science is organized.

In other words, like many things that may seem obvious to us once we grasp, understand and can articulate them, the components of critical science literacy are not necessarily obvious to everyone. Much social knowledge is tacit or implicit rather than formally learned. Ask a scientist—or a science journalist or science studies scholar—how they identify scientific truth as opposed to scientific error. They likely rely on their knowledge of the social dimensions of science (how science works, in other words), and they likely also rely on heuristic cues about what science (and what scientists) to trust. But they will not necessarily give this answer, and they may not even be fully aware of it. To many of us, science is a foreign culture, one for which there is no obvious opportunity to gain familiarity. Knowing a foreign culture and its expectations is often an implicit accomplishment, and it always requires exposure. While gaining familiarity in this sense may not always be a stated goal of the public engagement movement, it is one of its chief contributions.

Going Forward

At present, these observations about the nature of science literacy and the role of public engagement can only help us to comprehend why communicating climate change is such a challenge, as well as suggest some new dimensions about what people actually need to know about science— including its social character—to make informed choices, dimensions we should learn to pay more attention to as science communicators and communication scholars. These ideas still need further development but may ultimately help us be better science communicators, at least in the long term. But with the rapid pace at which climate change is occurring and the very slow pace of political and governmental response at both the national and global levels, it is unfortunately not clear that new approaches to science communication or science literacy are going to solve the problem quickly enough. There are obstacles other than limited public familiarity with science to be overcome, and viable action strategies are also needed.

Climate change policy in the United States seems to be a special case in that this is a period of intense and sometimes bitter political polarization within U.S. society and politics, the same dynamics that have made Congressional consensus, at least for the time being, seem like a thing of the past. For the first time in U.S. history, the 2016 presidential election race included, in addition to more conventional Republican and Democratic party choices, presidential candidates from the left, a democratic socialist and an environmental activist; from the right, a diverse array of religious, economic, and social conservatives; and from the far right, a businessman described by some commentators as a neofascist. Arguably, we have never had such a broad spectrum of political candidates for president, and it seems likely we could see a changed constellation of political parties emerging in future. Where climate will end up in terms of relative priority remains to be seen. And the political dynamics of getting the developed and "less developed" worlds to agree on who has to fix the problem offers another set of challenges of even substantially greater scale, one on which great progress was made at the 2015 Paris climate talks—but this progress only matters if individual nations follow through. In other words, public understanding of science is hardly the only factor inhibiting change.

Yet, over time, the proportion of Americans accepting climate change has crept up, and we believe that it will continue to do so. The collective dynamics of our highly diverse, pluralistic, multi-cultural society produces great differences of opinion, but in the past, consensus has ultimately emerged on many important issues despite this diversity. Consensus on climate is not likely to happen very quickly, especially during a period in which the U.S. political system is greatly in flux—some might even say it is in disarray. In seeking a path forward, we have chosen to focus more on the capacities of the strong majority of people (in the United States as in many other areas of the world) who accept climate change than on confronting minority "denier" perspectives. The majority who recognize and understand the basics of the problem are the ones most in a position to make real change.

Communication scholars, science communication practitioners, and others interested in promoting science literacy and climate change awareness should take note of how science actually works and communicate this to audiences. But this will definitely not be a short-term fix. Further, even those armed with deeper background knowledge of how science works will also be constrained by the political climate and by the agendas of media organizations. Given the weak (but emerging) NGO infrastructure dedicated to climate, voices powerful enough to influence that climate and those agendas are in short supply. Media usually give most of their attention to the statements of public figures and major contemporary events, not scientific developments or longer-term policy issues. Even in the elite press, newsworthy stories are generally those that are both novel and dramatic. Escalating climate disasters will ultimately fit this pattern, but we cannot wait that long. The next chapter will analyze the dynamics of the media attention cycle and begin to suggest how we can tip the balance.

Notes

1. Worth noting in this context is the proposition advanced by feminist scholars of science that the conduct of science is itself imbued with values and beliefs; see Fausto-Sterling (1987). Injecting values into science may be almost inevitable at times, but we seek to minimize it even so.
2. The idea that society is self-correcting is an important tenet of functionalism, the theoretical perspective adopted by Robert Merton in his work. The self-correcting capacity of the social system often seems limited, however, and this theoretical perspective has difficulty accounting for social change. Far fewer sociologists today describe themselves as functionalists, in large part for this reason.
3. An additional problem with analyzing trends in this kind of data on factual knowledge is that, just as for opinion poll questions, the context (including the emphases and wordings used in science and science education) can change over time, making apparent trends at least hypothetically reflective of contextual factors applying at different points in time rather than the changes we think we are measuring in understanding (see Bishop 2004).

References

American Institute of Physics. 2015. The Discovery of Global Warming. https://www.aip.org/history/climate/timeline.htm

Bhattacharjee, Y. 2005. Citizens Supplement Work of Cornell Researchers. *Science* 308(5727): 1402–1403.

Bishop, G. F. 2004. *The Illusion of Public Opinion: Fact and Artifact in American Public Opinion Polls.* Rowman and Littlefield.

Clarke, C.E. 2008. A Question of Balance: The Autism-Vaccine Controversy in the British and American Elite Press. *Science Communication* 30(1): 77–107.

Fausto-Sterling, A. 1987. Society Writes Biology/Biology Constructs Gender. *Daedulus* 116(4): 61–76.

Harris, G. 2010. British Journal Retracts Paper Linking Autism and Vaccines. *New York Times*, February 2. www.nytimes.com/2010/02/03/health/research/03lancet.html?_r=0

Kuhn, T. S. 1970. *The Structure of Scientific Revolutions.* 2nd ed. University of Chicago Press.

Latour, B., and S. Woolgar. 1986. *Laboratory Life: The Construction of Scientific Facts.* 2nd ed. Princeton University Press.

National Science Foundation. 2014. Science and Engineering Indicators 2014. http://www.nsf.gov/statistics/seind14/index.cfm/chapter-7/c7h.htm

Priest, S. 2013. Can Strategic and Democratic Goals Coexist in Communication Science? Nanotechnology as a Case Study in the Ethics of Science Communication and the Need for "Critical" Science Literacy. In *Ethical Issues in Science Communication: A Theory-Based Approach*, eds. J. Goodwin, M. F. Dahlstrom, and S. Priest, 229–244. Proceedings of the Third Summer Symposium on Science Communication, Iowa State University, May 30–June 1.

Rughinis, C. 2011. A Lucky Answer to a Fair Question: Conceptual, Methodological, and Moral Implications of Including Items on Human Evolution in Scientific Literacy Surveys. *Science Communication* 33(4): 501–532.

Science/American Association for the Advancement of Science. 2016. Special Online Collection: Hwang et al. Controversy—Committee Report, Response, and Background. http://www.sciencemag.org/site/feature/misc/webfeat/hwang2005/

Secko, D., E. Amend, and T. Friday. 2013. Four Models of Science Journalism: A Synthesis and Practical Assessment. *Journalism Practice* 7(1): 62–80.

Seethaler, S. 2016. Shades of Grey in Vaccination Decision Making: Tradeoffs, Heuristics, and Implications. *Science Communication* 38(2): 261–271.

Spence, W., R. B. Herrmann, A. C. Johnston, and G. Reagor. 1993. *U.S. Geological Survey Circular 1083. Responses to Iben Browning's Prediction of a 1990 New Madrid, Missouri, earthquake.* U.S. Government Printing Office. http://pubs.usgs.gov/circ/1993/1083/report.pdf

Sturgis, P., and N. Allum. 2004. Science in Society: Re-Evaluating the Deficit Model. *Public Understanding of Science* 13(1): 55–74.

Wynne, B. 1989. Sheepfarming After Chernobyl: A Case Study in Communicating Scientific Information. *Environment* 31(2): 10–39.

CHAPTER 7

Ingredients of a Successful Climate Movement

Much of this chapter is based (with updates and extensions) on an earlier paper by Neil Stenhouse and Susanna Priest that was presented at the Conference on Communication and the Environment (COCE) in Uppsala, Sweden, in 2013.

Previous chapters have considered the structure and effectiveness of persuasive climate messages, a variety of factors affecting people's receptivity to them in a complex opinion climate, how information consumers find and process climate information in a changed news environment, and why weak background knowledge of the social organization of science ("how science works", Chap. 6) can limit both citizens' and journalists' capacity to assess scientific claims. Here we address another media problem—the factors that create the attention cycle for news—and the type of bona fide social movement that will likely be required to overcome it.

The reliance on "false balance" in news stories, resulting in the awarding of legitimacy (even if inadvertently) to "scientific" claims that do not reflect the current scientific consensus, is not the only problem in media coverage of climate. Social context, both how science matters to the lives of non-scientists and the social character of science itself, is often missing from media accounts. And generating sustained media attention can be difficult to achieve. Setting the agenda for public discussion (a phenomenon recognized by McCombs and Shaw in their now-classic 1972 article on this topic) is one of the most familiar and best-documented media effects. If there is no media coverage of an issue that most people do not directly confront in other ways, attention recedes as though the issue hardly exists—or at least is unimportant. If there is coverage, regardless of the tone or other content details, people tend to think the issue is more important—it rises on their personal agenda.

Agenda-setting in itself is not always defined as a problem; it simply describes one of the more powerful consequences of the actions taken by journalists and editors, in their roles as news gatekeepers, to choose particular issues and events to feature. However, an implication of this observable pattern—its converse, actually—is that if there are no novel developments seen as newsworthy, then media attention is likely to drop off, and the issue will slip down on the personal agenda of news consumers. Lack of attention is rarely studied by scholars; it leaves no artifacts in the form of published articles for researchers to work from. But we can reasonably assume that diminished coverage will result in diminished attention from both the public and the political sphere. This phenomenon does not derive solely from journalists' judging (or misjudging) what issues are important, or even from the quest to sell news. It stems from the fact that much news, especially science news, is generated as a result of press releases and other information subsidies.

Once attention fades, journalism (of all types) tends to ignore the current problem and move onto the next. In his influential 1972 article, Anthony Downs proposed that social issues possessing certain characteristics tend to pass through an "issue-attention cycle" of five phases. In the first or "pre-problem" phase, the social conditions that will come to define the issue exist, but are noticed and thought about by a few experts at most. In the second phase, the issue is explosively "discovered" by the public (via the news media), and calls are made for "something to be done". In the third phase, as solutions are sought and legislative options are considered, people begin to discover that the problem will not be so easy or cheap to fix as had previously been imagined; attention levels off as solutions begin to seem elusive. In the fourth phase, due to increasing recognition of the difficulty and true cost of solving the issue as well as the unpleasantness of even acknowledging its continued existence, public and media attention begins to wane. In the final or fifth "post-problem" phase, the issue enters what Downs called a "prolonged limbo" (1972: p. 21) in which public attention is low, but due to the issue having already passed through the cycle and being available to collective memory, attention may flare up again in response to some new circumstance more rapidly than for an entirely new issue.

Downs also suggested that there were three characteristics that defined the types of issues that would pass through this cycle (1972). The first is that the number of people affected by the issue must be reasonably large in absolute terms, but fairly small as a proportion of the total population. This would make the issue seem big enough to be worthy of national attention, yet small enough that most people will not have their attention kept fixed to it by personal experience of the problem. The second characteristic is that some aspects of the social arrangements that cause the problem also bring significant benefits to a majority of people, or at least to some powerful minority. This would make for a significant level of disincentive towards actually solving the problem—somebody would lose out from a solution. The third characteristic is that the issue is not intrinsically exciting. This would mean that the stories would not be appealing to broad mass audiences (or at least are not seen as likely to be appealing by news editors) and would quickly drop off of the public agenda, either because the public stopped reading or watching the reports that did exist, or because journalists' expectations of public disinterest results in a shift of reporting to other topics (in other words, journalists might quit seeing the issue as a "story"). Below, we explore what this analysis suggests about the issue of climate change.

Gaining Attention for Climate

Climate change fits all three of Downs' criteria in some ways. Ultimately, everyone on Earth will be affected; however, the proportion of the U.S. population that will be directly, significantly, and visibly affected in the short term—among others, people who live on ocean beaches at low elevations, or on rapidly eroding bluffs, and farmers whose agricultural cycle is being affected, for example—is likely to remain relatively small, at least for the next few years (AAAS Climate Science Panel 2014). Others may experience such things as local drought, stronger storms or more common threats from wildfires, but the connections to general climate trends may seem less certain. Meanwhile, all U.S. citizens (like many others around the world) benefit massively from the primary cause of the problem, the combustion of fossil fuels (Anderegg et al. 2010). In addition, a very powerful set of interests within society, the fossil fuel industry, is directly threatened by the prospect of climate change-related regulation—or at least is likely to think that it is threatened. And the issue itself may not be seen as intrinsically exciting or threatening. A number of apocalyptic climate-related disaster films have certainly been produced, but it is easy to imagine them being dismissed as factitious, exaggerated and sensationalistic.

The most dramatic impacts of climate change, those that we can imagine would otherwise draw attention, are either seen as uncertain to occur or (as in the case of extreme weather) their causal link to climate change is difficult to present in a way that is intuitively compelling (Wynn 2012; Leiserowitz et al. 2012). These impacts may also be imagined to involve places that are distant and people that are strangers, rather than family, friends and self. And the misperception that we can wait many years before taking action (since the worst effects may indeed take many years) also appears to be common, resulting in problems seen as more immediately pressing taking priority.

McComas and Shanahan described the first "pass" of climate change through the attention cycle (1999). Climate change has recently made at least one more pass through the cycle. In this second pass, following a new "pre-problem" period of inattention outside of expert circles, the surge of publicity associated with the release of the movie *An Inconvenient Truth*, the release of the International Panel on Climate Change (IPCC) Fourth Assessment Report, and the awarding of the Nobel Prize to the IPCC and Al Gore constitute the second or "discovery" phase (Mooney 2011). The convening and then the breakdown of the 2009 Copenhagen talks and the failure of the climate bill in the U.S. Senate in 2010 characterize the third or "discovery of significant costs" phase this time around (Mooney 2011). And the subsequent decline in media attention and downgrading of the perceived seriousness of the problem fits the fourth phase (Brulle et al. 2012). Recent evidence that this decline in attention may have stopped is consistent with the "post-problem" stage (Borick and Rabe 2012).

In 2015, with the release of a new IPCC report (the Fifth Assessment Report) and the encouraging progress made at the Conference of the Parties or COP meeting in Paris[1] resulting in a major commitment to "nationally determined contributions" to combating climate change, we certainly seemed to be entering another of these cycles. But present circumstances highlight an additional dynamic affecting this cycle: the size of the so-called "news hole" and the de facto limitations on capacity in both news coverage and public attention. The term "news hole" was more common when newspapers and a handful of broadcast networks were the mainstay news media for most of the population; it refers to the fact that only a limited amount of space (the "news hole") is available for actual "hard" (timely, event-driven) news stories, after allowing allocations for advertising and other kinds of routine content (recipes and advice columns, the weather, "softer" features). In print media this is still very much the case, especially considering the space available on the front page and in the first section, and in broadcast media, the "space"—or, rather, time—may be even more limited; on the new "24/7" news cycle in the United States created by the emergence of several all-news, all-the-time networks (notably Fox, CNN, and MSNBC), there seems to be no limit—but there is a lot of repetition. Major mainstream media, whether print, broadcast or web-based, tend to focus on a limited number of stories at a time. And during 2016, in the United States the central story was the presidential campaign, which pushed almost everything else aside—including climate.[2] When this has faded, other issues will compete for the attention.

News media tend to imitate one another with respect to choices about what is the most important story of the moment, a phenomenon sometimes referred to as "inter-media agenda setting" (see, e.g., Sweetser et al. 2008). This tends to reduce the range of issues readily available to individual news consumers (especially those not actively seeking specific issue information, for example on the Internet, in line with their own interests and priorities). In addition, the available attention among news consumers might be thought of as a limited commodity in itself. In Europe, media attention as of this writing is focused on events such as the Syrian refugee crisis and the significant terrorist attacks that occurred in Paris in November of 2015, just prior to the COP 21 meeting, and in Brussels in late March of 2016. In the United States, recent attention has been riveted on the historic and often bitter competition among prospective presidential candidates—capturing the attention both of the politically engaged and those normally unengaged and disinterested. Climate? We can deal with that later, can't we? And anyway, do we really have to?

Yet the issue-attention cycle can be broken—or at least re-booted. Startling new findings from climate science, notably James Hansen and colleagues' recent claim that sea levels are rising dramatically faster than anticipated (2016), still get significant press attention, especially when proactively publicized, in both mainstream media and the blogosphere. (The Hansen announcement, which predicted the loss of major coastal cities worldwide, wisely included an accompanying video.) Protests also tend to get attention: On March 23, 2016, several hundred protesters marched into the Superdome in New Orleans, made iconic a decade earlier as the woefully inadequate temporary shelter for many of those fleeing Hurricane Katrina in 2005. CNN—among others—covered the recent protest, an attempt to stop the leasing of 45 million Gulf of Mexico acres for "fossil fuel development". But the CNN discussion, in the form of an op-ed piece by John Sutter (2016), described the scene as "wild—often totally excessive", and Sutter said that it "mystified" him,[3] while the oil and gas representatives who had come there to bid on the leasing rights were described as looking on "quizzically". Just as the media can confer legitimacy, it can also delegitimize dissent (Gitlin 1980).

This illustrates the problem that the kind of drama that can break this cycle can also be used to make both climate change research and climate scientists seem less than legitimate—a classic two-edged sword. Indeed, after the Hansen results were published in an open access, open review journal (meaning that reviewer comments are transparently available), one commentator counted them and found that of the 60 comments, an "exceptional number" for an open review, one third were from apparent climate deniers. Yet the title of her article for the Climate Home website called the study "apocalyptic" and the subtitle proclaimed that "many heavyweight critiques" had accused Hansen of "unprofessional behavior" and alarmism (Darby 2016). Just one critic making the "unprofessional behavior" claim is actually quoted in the article, although milder criticisms from several other sources are also offered—making the subtitle seem rather misleading. (The web article discusses the underlying scientific contribution of the paper as well, but not until halfway through it.) For those unfamiliar with the news business, it is worth pointing out that headlines are often written by a news editor, not the journalist writing the story, and it seems reasonable to suppose that this tradition has been carried forward into the website world. But regardless of who is responsible or whether Climate Home wanted to undermine Hansen's work or simply to gain readership, the likely effect is obvious: both of the above.

Given that even dramatic new findings such as Hansen's cannot be relied upon to make a persistent disruption of the issue-attention cycle, a more sustained compelling force will likely be needed to turn the tide on climate change action. A broader social movement is one phenomenon that may fit this description. Few communication scholars have explored the idea of a climate movement (as opposed to a more general environmental movement), although more are beginning to do so in recent years (e.g., Endres et al. 2009; Han and Stenhouse 2015; Hestres 2015; Nisbet 2014; Pearson and Schuldt 2014). Even fewer have considered the characteristics of successful movements that have had sustained impacts on society. These collective activities are an important, even an institutionalized, component of modern life in a democracy. As communication scholars and others interested in successful climate communication strategies, we should know more about them—and as communication researchers, we might devote more attention to the role of communication in their rise and fall.

What Scholarship Tells Us About Social Movements

After the failure of the 2009 talks in Copenhagen to generate a binding global deal and the failure of the 2010 climate bill in the U.S. Senate, many people wondered what had gone wrong with the push toward strong climate policy and what to do next (Walsh 2011). Several explanations have been put forward for why the issue did not continue to receive high levels of attention. One of the explanations that has gained some adherents is that the will for action on climate change, despite receiving majority support in most polls, is not organized or deep or serious enough (Roberts 2011; Palmer 2012; Skocpol 2013). Proponents of this view argue that there needs to be a stronger social movement on climate change, similar to the civil rights or suffrage movements, or to the earlier environmental movement of the 1970s (Amenta et al. 2010). However, the idea of a climate movement—rather than a general environmental movement—has been little studied in communication scholarship (for a notable exception, see Endres et al. 2009). Contributing factors to the lack of attention to global warming and low ratings of global warming as a national priority include the low levels of public mobilization and the low visibility of any kind of broad-based grassroots social movement for climate change (Romm 2011; Roberts 2011; Brulle 2010).

Part of the problem appears to be that climate change does not neatly match the agenda of most existing organized groups such as the major U.S. political parties or even existing environmental activist groups (see Chap. 5 of this book). These often focus on more specific objectives such as conservation of wildlife and wilderness, stemming specific sources of toxic pollution, or halting practices that seem to carry especially high levels of short-term environmental risk, such as nuclear or hydroelectric power development. This is important work. Ironically, though, these same high-risk power alternatives to fossil fuel that may be anathema to many U.S. environmentalists can seen by others as part of the path forward to combat climate change. Accelerating development of natural gas (a "cleaner" fossil fuel), as proposed in President Obama's 2013 State of the Union address, could be even less popular with some environmentalists. A climate change movement has interests that do not always align neatly with those of the existing environmental movement in the United States; the issue of climate could need its own movement with its own organizational infrastructure.

A well-organized and broad social movement for climate could help to maintain public and media attention on the issue, and thus increase the likelihood of the issue staying on the political agenda and strong climate policies getting passed. One of the most important consequences of social movements may well be their impact on the political and media agenda. The historical record shows many examples of social movement activity helping bring attention to an issue by way of direct person-to-person consciousness raising, through stunts that gain media attention, via rallies and other traditional protest activity, by issuance of provocative statements and letters to the editor, by the organization of pressure campaigns on lawmakers who then address the issue publicly, and by other methods. Climate movement actions have proven they can help gain media attention (as happened in the case of protest in the United States against construction of the next expansion, known as Keystone XL, of a Texas-to-Alberta oil pipeline, a project that was subsequently halted by presidential action). The involvement of prominent climate scientists in movement activity enhances these organizations' legitimacy and also provides journalists with more to write about.

A social movement organized around climate also has the potential to help overcome two of the characteristics that Downs suggested allow issues to recede from attention: the number of people who perceive themselves as affected and the interest people have in the issue. And an active enough minority should be able to maintain attention, even to causes that are not seen as affecting the majority of people, through effective organization and media "savvy".[4] Additionally, by working to find and highlight the most interesting dimensions of the issue and by creating dramatic news events for media to respond to, a movement organization can provide "news hooks" that facilitate coverage by busy reporters and increase media, and thus potentially policy maker, attention. Organizations that provide the media with news and information get their issues—and their sides of those issues—covered; "information subsidies" can work as well for them as for corporate and other interests.

A climate-specific social movement holds potential promise for sustained activity that can maintain media and political attention to climate change, a critical function to keep attention on this "slow-moving disaster". However, in order to be successful, this activity will need to be maintained over decades (Smil 2010). This will be difficult. For this reason, a successful climate movement will benefit from considering the factors that have been shown to be relevant to movement success in the past. In the following sections, we describe four general concepts from the social movement literature that reveal insights about what a climate movement would need to do to succeed in the long term. These involve movements' access to resources (resource mobilization), their ideological makeup (collective identity), the political environment in which they find themselves (political opportunity structure), and the strategies they choose to make their way in the world, taking into account the other three factors (strategy and strategic capacity).[5] We also highlight the potential contributions of communication and communication research.

Resource Mobilization

A major insight of the resource mobilization literature is the idea that, while of course social movements need resources to succeed, these resources can take many forms—people, money, skilled personnel, experience, social status, connections, and so on. A successful movement will need to mobilize—to organize and deploy—at least one of these resources in order to be successful. However, movements need not possess vast amounts of all resources. Social movements can succeed in many ways—by having large numbers of moderately committed members, by having a small but intensely committed and organized base, or by being well connected to community leaders or political elites. They do not all need to be giants in order to be successful (see, e.g., McCarthy and Zald 1977; Edwards and McCarthy 2004).

Given that environmental protection has become an established, accepted, reasonably mainstream goal, some of the established environmental organizations have significant income from multiple donors, and they sometimes have good connections to lawmakers and other elites, but they may have less of an ongoing direct connection to much of their membership base. In the context of climate legislation, this means that when elite negotiations fail, these institutionalized, established environmental organizations may find it difficult to rally their membership and have them exert political pressure. When the relationship with an organization consists of a regular check, the member becomes happy to give that check, but arguably less happy to get involved in effortful political action. This kind of pattern was predicted by McCarthy and Zald (1977). Established organizations may have good resources in terms of finances and elite connections, yet still wield little real grassroots power even where membership numbers may be large.

A solution might involve mobilization of a different resource instead of, or in addition to, the broad but shallow memberships supported by insider lobbying characteristic of established movement organizations. Both true local grassroots memberships and active protest may be required. There is substantial evidence of this kind of strategy being successful and affecting legislation in the past (Olzak and Ryo 2007). So why is this not happening more for climate? Some observers have characterized the United States as a nation of disconnected, disengaged individuals, epitomized by Robert Putnam's discussion of the erosion of social capital in *Bowling Alone* (2000). The proliferation of individualized media, "social" or not, can hardly have slowed this trend and may even have accelerated it—in effect making U.S. citizens more passive.

Yet the scholarly literature also indicates that active mobilization on a large scale can play a significant role in achieving a movement's goals. A recent review of articles on 54 social movements in the United States found that the larger movements were substantially more likely to have an effect (Amenta et al. 2010). Other evidence has also suggested that the size of a movement—measured in terms of number of organizations rather than number of individuals—may have an indirect effect on success by increasing the diversity of tactics used by the groups. Olzak and Ryo (2007) studied black civil rights organizations that had operated during the U.S. civil rights movement. They found that as the number of organizations increased, the diversity of tactics—such as lobbying, boycotts and marches—tended to increase, and increased diversity of tactics tended to be followed by increased success for the movement. However, as the number of organizations increased beyond 100, both the diversity of tactics and levels of support were reduced. Movement success may not be determined by movement size alone, in other words.

The resource mobilization literature helps us to visualize the possibility of a climate movement by providing various models of how a complex organization (or group of organizations) can be constituted and demonstrating that its success will largely depend on how well it acquires and deploys many different types of resources. Whether U.S. society today is organized in a way that can and will facilitate this for a problem as diffuse, abstract and seemingly distant as addressing climate change is another question. We still have much still to learn about the social ecology of social movements—and about what kind of communication activity is optimal for the accrual of a diverse range of essential resources, including various forms of membership mobilization. Strategic communication specialists should be able to make a contribution to this effort.

Collective Identity

The term collective identity has been used in several ways by scholars. Generally speaking, it refers to a group's sense (or individual members' senses) of who they are as a group—what it means, in terms of identity, to belong, in other words (Ashmore et al. 2004). Collective identity in the sense of how different groups perceive themselves—and are perceived by others—as distinct can play a large role in determining which individuals will join a movement, as well as whether they continue to commit time and resources to it after joining. Evidence from the civil rights movement (and others) has shown that people are more likely to join particular movement organizations if they share strong ties with others who are members or if they feel a strong attachment to the ideals expressed by the movement.

In this way, collective identity can serve as one solution to the "free-rider" problem of leaving solutions to others, explaining why a rational person would bother to make extra effort when success is indeterminate or unlikely and all people will share the benefits if the movement is successful (Polletta and Jasper 2001). Indeed, non-blacks who participated in the U.S. civil rights movement likely gained exactly this sense of shared benefits, if only "feel good" benefits, from striving for a more just society. Having a distinctive narrative can help an organization maintain a distinct identity, justify its purpose, and help maintain continued commitment.

Collective identity is also important because it affects the activities that movements undertake. If a movement has a radical identity, for example, its members are more likely to be opposed to collusion with corporate power; if a movement has a moderate identity, its members are less likely to want to protest—or get arrested. Even if the circumstances suggest such counter-identity moves could bring significant returns, movement members would be reluctant to engage in them since it would go against their sense of who they are as a group. Collective identity can outweigh purely materialistic cost-benefit calculations. It is not a static entity that rigidly enables certain actions and limits others, however. A particular "external" identity can be cultivated in order to obtain certain benefits, such as respect. And in situations where certain activities that seem to be desirable would force members to step outside the current identity, the group can alter their identity in order to make the new actions more possible and more plausible.

Movements can also use their collective identity strategically and deliberately, in order to elicit specific reactions from others or to make it more difficult for others to use specific tactics against them. A movement that refers to itself as "the people" (as opposed to embracing a narrower identity) may get greater numbers of individuals to participate; a moderate environmental organization may use its history of cooperation with corporations to create a moderate identity and prevent opponents from attacking them as extremist radicals. This suggests that a range of organizations with multiple identities filling different niches should be successful. Importantly, though, collective identity also limits the kinds of alliances that can be made by movement leaders without affecting membership. If movement members join because an organization's collective identity resonates with them, and leaders then decide to make an alliance that members perceived to be inconsistent with that identity, they may leave. Climate change represents a particular challenge in this regard because of the eventual need for both environmental groups and energy interests to participate in implementing solutions.

Collective identity is thus much more important than simply defining the wrapping that movements choose to cover themselves in. It has important material and strategic consequences also. As well as being vital in enabling movement organizations to exist, to draw members and to maintain action, collective identity has important effects on the types of actions that are available to movements and to the strategic choices they have to make. Collective identity would certainly be an essential element within any effective climate movement group, but this will vary with the type of organization. While those involved with one organization might identify primarily as climate-oriented activists, others may see themselves as part of a broader environmental or social justice movement, with their climate-related activities flowing from those broader goals. Yet others working in energy industries or in energy policy may share many of the same goals and values as those in the climate movement, but may not see themselves as part of a "movement" at all—indeed, some might be restrained for fear of being seen as the "wrong kind" of person (a type of "spiral of silence" effect), although others may see a potential synergy instead of a conflict.

Despite these complexities, a strong, overarching, shared collective identity may help encourage more people and more groups to work together. There will be opportunities for productive alliances across groups, such as between environmental and energy interests, by adapting *both* identities to embrace broad but sustainable industrial development, as well as smaller-scale commercial efforts that experiment with alternatives on a more local level. The term "sustainable" has itself has become a buzzword because it is a concept that people and groups with highly divergent identities, working at various scales toward different but generally compatible goals, can all accept—even identify with.

These kinds of alliances may not be attractive for some, and it may be better for some groups to work more independently, maintaining an "outsider", "fringe", or "pure protest" identity. Very likely there are limits to how much particular groups' identities can be shifted. In some cases broadening an organizational identity may result in a reduction of passion and identification with the group. More moderate participants may react by dropping out of an organization they see as increasingly radicalized—or, conversely, some participants may drop out of an organization they believe has been co-opted by becoming too mainstream.

Clearly, the development, cultivation, projection, and modification of organizational identity should be an area where communicators and communication researchers can make a major contribution.

Political Opportunity Structure

Another way in which movements can enhance their effect is by aligning themselves with the existing political opportunity structure (Meyer and Minkoff 2004). Broadly speaking, this phrase refers to the political circumstances in which a movement seeks to act. While mobilization and collective identity focus on the internal activities and makeup of a movement, political opportunity structure focuses on the external environment in which the movement will be acting. Important factors in this environment include, for example, the nature of the political system (including whether local or national), the amount and type of power available to the different political parties and other relevant groups, the resources available to be mobilized, specific legislative rules, recent political history, and the strength of existing alliances (Meyer and Minkoff 2004).

One relevant early work in this domain was a 1986 study by Kitschelt comparing the history and effectiveness of anti-nuclear movements in

France, Sweden, the United States, and West Germany. The dimensions Kitschelt looked at were political input structures—that is, how easy it was for new issues and agendas to be introduced for serious consideration—and political output structures—that is, how easy it was for new agendas, having entered the arena, to establish an effective coalition of allies and be established in policy.

For example, at the time of the study, the United States was classified as having "open" input structures, due to the relative strength of Congress and the fragmentation of the executive branch, and "weak" output structures, due to these same factors and to the overall lack of an official structured process for interest groups to interact with government. (Today, this same characterization of the United States might not apply, with a gridlocked Congress and an arguably stronger executive branch.) France, conversely, had "closed" input structures, due to its strong executive branch keeping attention fixed mainly on issues important to the two main political blocs, and "strong" output structures, due to the executive arm facing few challenges to any moves to implement the policy that it decides to make.

Kitschelt concluded that these patterns of political opportunity structure resulted in anti-nuclear movements engaging in more lobbying in the United States, where input structures were more open to direct influence, and less lobbying in France, where input structures were closed. Protests against nuclear power were relatively larger in France, since more "insider" strategies were closed off to the French anti-nuclear movement; protests in the United States were relatively smaller, and they only achieved large numbers in the wake of major incidents such as the partial meltdown at Three Mile Island (Kitschelt 1986).

More recent studies in the opportunity structure literature have criticized studies like Kitschelt's as being overly simplistic and have shown how important variations in opportunity structure are much more intricate and variable than closed/open input and weak/strong output (Amenta et al. 2010). For example, the output structures in Italy can be seen as having been weaker for environmental policy than other policy areas until a Ministry for the Environment was set up (Giugni 2004). However, this only strengthens the argument that significant attention to the opportunity structures in a particular country at a particular time is warranted for any movement that wishes to achieve recognition and success within that country.

Future attempts at mobilization ought to take into account the specific opportunity structures relevant to the goals of the movement if they want to improve their chances of success. Those promoting change should attempt to achieve goals that are possible in the political environment in which they are acting, as well as take what steps might change that environment for the future. Political communication scholars, among others, should find opportunities here to study how political opportunity structures limit or enhance the effectiveness of particular communication strategies.

Strategic Capacity

Scholars of strategy as used in social movements have noted that it is difficult to create generalizable rules (Maney et al. 2012). Each case of strategic planning is a complex process requiring the consideration of multiple possible futures and tradeoffs between different choices, the outcomes of which are usually impossible to know with certainty. The appropriateness of a strategy depends to a large degree on the particular circumstances and necessarily incomplete information available at the time at which the strategy is constructed, and a good strategy in the face of one particular set of circumstances may be inappropriate or inapplicable at another particular time. Indeed, one important part of strategy is the ability to change course rapidly in response to changed circumstances, rather than staying fixed to a rigid plan.

While it may thus be impossible to create a blueprint for effective strategy that is applicable to all social movements, Ganz (2004) has devoted significant attention to the generalizable characteristics of social movement organizations that may help increase the chances of those organizations making good strategic choices. Ganz describes various elements that can add to a movement's strategic capacity. These include members with a deep knowledge of the issue the movement is based around, including its entire social and political context; an organizational makeup that results in members regularly encountering diverse perspectives on the movement issue; organizational processes that facilitate the expression of minority views; leadership consisting of both "insiders" with deep knowledge and experience with the issue and "outsiders" who can use their unconventional perspectives to suggest innovative responses to situations; and a financial and physical resource base drawn from multiple sources, so that the organization is not constrained in their action by the need to maintain their appeal to any one particular perspective on the issue. This work on strategic capacity is consistent with the results of Olzak and Ryo's study (2007) showing that greater diversity of tactics was beneficial to a movement.

Strategic capacity is an organizational characteristic that the work of these scholars suggests is enhanced by a strong knowledge base and diverse leadership providing diverse points of view. Organizational communication specialists may find this perspective could suggest interesting opportunities for research on how communication among diverse organizational constituents supports strategic decision-making. Science communication specialists in particular should be able to add value to efforts to consolidate background knowledge and make it accessible to both leaders and members.

Lessons and Opportunities for Communication

Although there are substantial differences between climate change and other issues that have inspired and mobilized social movements in the past, much can be learned from studying the important aspects of past social movements and applying them to the present. In the above discussion, we have tried to suggest how particular branches of communication scholarship could contribute in each of the areas that have been identified as important for the success of movement organizations. In this concluding section, we elaborate further on some of the applied research opportunities for communication scholars suggested by the four areas of social movement literature we have outlined. These observations should also be of interest to communication practitioners wanting to address issues of practical relevance to current movement organizations—or as "food for thought" for anyone attempting to mobilize collective action on climate change.

The resource mobilization perspective suggests that those wishing to promote collective action on climate should find ways to promote the particular types of mobilization that a climate movement or similar organization is likely to require for their goals to be achieved. Individuals must not merely understand the problem of climate change and have positive attitudes towards the movement and its purposes. Although certainly beneficial, a positive impression is a necessary but not sufficient condition to ensure actual participation. The goal is different from the goal of changing isolated behaviors, such as buying a different product or adopting healthier behaviors, goals to which so much research in communication has been devoted. Nor is the goal simply to persuade people of the validity of relevant scientific facts, a "deficit" approach. Rather, people must be motivated to donate their time, effort, money, skills, social contacts, or other resources to the movement. But what kinds of messages will motivate these kinds of choices? Some scholarship has suggested "framing for social action" as opposed to "framing for persuasion" as a productive way to think about this kind of strategic communication goal (Sprain 2015). Promoting mobilization—framing for social action—will require changing the types of message appeals used. More research on exactly what this will look like is certainly required and should add to our more general understanding of persuasion by presenting a different kind of case that has received relatively less attention from researchers.

The role of collective identity should prove to be extremely important in communicating to potential participants, constituting a sort of "branding" of the organization and its purposes. The sense of positive collective identity (and self-identity) that comes with movement participation can serve in lieu of material rewards in encouraging people to make the efforts involved with participation. Appeals to values such as preserving wilderness and other animal habitat or creating a healthier, safer, and more secure world for human beings should certainly be powerful in shaping attitudes and even encouraging membership. Values matter; appeals to the money to be saved by making individual decisions to conserve energy or adopt residential alternative energy options like solar or wind power are less effective with people who are not also motivated by environmental values (Priest et al. 2015). But appealing to collective identity should evoke an even deeper sense of self and character and enhance motivation to become actively involved.

Yet few communication scholars have examined the relationship between movement appeals and collective identity—that is, what participation in a particular type of social movement says about who the participants are, what identity an organization needs to adopt to attract a specific kind of member, and how participation changes people's identities and then considering how this would translate to climate. Examining the nature, importance, and effectiveness of appeals to collective identity in eliciting and sustaining participation in past movements should be a fruitful direction for future research.

Individual organizations should also consciously and carefully consider what kind of collective identity they wish to project and exactly how this will affect the amount and types of participation they receive. This is not to say that an organization should always aim to maximize the number of possible members and alliances, which might dilute their appeal and thus their eventual impact. Rather, those wishing for action should think carefully about whose participation is likely to be crucial and effective, and even whose allegiance may be unnecessary or harmful, before making decisions that affect their collective identity and its public expression.

Considerations of political opportunity structure suggest that a climate movement should aim to have its effects via the leverage points in the political system through which the movement is likely to be able to influence outcomes, as well as where those outcomes will have a big enough effect on carbon and other greenhouse gas emissions to be worth the effort. One of the leverage points with perhaps the largest effect on U.S. emissions would be the federal government. However, Skocpol (2013) argues that the climate movement is unlikely to have a strong effect there in the short term, due to the entrenched anti-regulation stances of both Senate Republicans and coal state Democrats. It might make more sense in the short term to focus on climate policy and clean energy development at a more local level. This could involve developing messages that invoke region-, state- or city-specific identities, values, key industries and occupations, and shared history, as well as available development and regulatory opportunities. Successful messages tailored to local conditions could help build local grassroots movements that could in turn help to elect pro-climate legislators who would help to build political pressure at the federal level.

Maximizing strategic capacity will require movement organizations to be flexible and adaptable. A successful climate organization will need to be able to react quickly to changing circumstances and emerging opportunities in order to pursue its strategic objectives most effectively. This suggests that applied communication scholars involved in the movement should be prepared to facilitate the development of effective new messages appropriate to circumstances quite rapidly, perhaps creating a set of "best practices" for this purpose. This would enable movement members to respond more quickly to unexpected extreme weather events that create teachable moments, sudden changes in legislation that create or close off mitigation paths, or signals that another actor's strategy (whether an opponent or an ally) is shifting.

Another aspect of developing strategic capacity with which communication specialists (both researchers and practitioners) can be helpful is enhancing access to the salient background knowledge available to movement members. If more people active in an organization have a greater amount of knowledge of relevant topic areas, this is likely to increase the chances of good decisions being made, as more potentially relevant factors will be considered from a greater variety of perspectives. Communication specialists could help improve access to this knowledge—both among movement members and among those the movement hopes to influence—by designing effective tools for storing and sharing relevant information on climate science, energy policy and alternative energy technology, political opportunities and constraints, good communication techniques, effective conservation strategies, and many other topics. This could be achieved in the form of a well-designed website and the development of talking points that could inform meetings, presentations, and other contacts within the group and between the group and important others.

Finally, having members with diverse perspectives is an important aspect of strategic capacity. But this could require that movement members with divergent perspectives have training in how to share views without causing offense—and leaders might be trained to best manage differences of opinion and other active conflicts as they arise. Communication specialists could help to suggest how communication opportunities could be designed so that members with diverse perspectives could share them in a way that best broadens the strategic perspective without detracting from an overall spirit of camaraderie.

Our final chapter will conclude this discussion by summarizing some of the important points made throughout this book and focusing on a few of what we consider the most important take-home ideas for communicators, communication researchers, and scientists.

Notes

1. Commonly referred to as the Paris Climate Change Conference, this meeting brings together the 195 countries that have ratified the United Nations Framework on Climate Change Convention (UNFCCC), which are referred to as the Convention's "Parties"; thus their meetings are referred to as Conferences of the Parties. The UNFCCC itself was adopted at the Rio Earth Summit in 1992. The commitments associated with the earlier Kyoto Protocol had expired in 2012. (See http://unfccc.int/essential_background/convention/items/6036.php for further details).
2. A google search conducted on March 25, 2016, for "climate protests 2016" returned 8.3 million hits, while "trump protests 2016" returned 95.8 million.
3. In fairness, Sutter also expressed his thanks to the protestors for calling his attention to U.S. federal leasing—for making the auction "visible". Some would argue that a little "wildness" would have been needed to make that happen, and even then coverage of this event was not widespread and seemed to fade quickly.

4. Three cases in point: About five times as many U.S. adults want *more* regulation of gun sales (55 %) as want *less* regulation (11 %; Swift 2015). However, it appears that the anti-regulation forces are far better organized and more vocal. Only 19 % of Americans oppose abortion under all circumstances (Saad 2015), and yet the U.S. healthcare system seems under constant attack for supporting the other 81 %. The contemporary Tea Party movement in the United States also suggests that an extremely committed minority can win out over a much larger, yet only moderately active, group (Oliver and Marwell 1988; Skocpol and Williamson 2011).
5. The literature in sociology and political science on social movements is vast and has important theoretical implications for understanding how society works more generally. Here, we attempt only to suggest a few of this literature's key themes.

References

AAAS Climate Science Panel. 2014. What We Know: The Reality, Risks and Response to Climate Change. American Association for the Advancement of Science. http://whatweknow.aaas.org/wp-content/uploads/2014/07/whatweknow_website.pdf

Amenta, E., N. Caren, E. Chiarello, and Y. Su. 2010. The Political Consequences of Social Movements. *Annual Review of Sociology* 36(1): 287–307.

Anderegg, W.R.L., J.W. Prall, J. Harold, and S.H. Schneider. 2010. Expert Credibility in Climate Change. *Proceedings of the National Academy of Sciences* 107(27): 12107–12109.

Ashmore, R.D., K. Deaux, and T. McLaughlin-Volpe. 2004. An Organizing Framework for Collective Identity: Articulation and Significance of Multidimensionality. *Psychological Bulletin* 130(1): 80–114.

Borick, C., and B. Rabe. 2012. Belief in Global Warming on the Rebound: Fall 2011 National Survey of American Public Opinion on Climate Change. The Brookings Institution. http://www.brookings.edu/research/papers/2012/02/climate-change-rabe-borick

Brulle, R.J. 2010. From Environmental Campaigns to Advancing the Public Dialog: Environmental Communication for Civic Engagement. *Environmental Communication: A Journal of Nature and Culture* 4(1): 82–98.

Brulle, R.J., J. Carmichael, and J.C. Jenkins. 2012. Shifting Public Opinion on Climate Change: An Empirical Assessment of Factors Influencing Concern Over Climate Change in the US, 2002–2010. *Climatic Change* 114(2): 169–188.

Darby, M. 2016. James Hansen's Apocalyptic Sea Level Study Lands to Mixed Reviews. Climate Home website. http://www.climatechangenews.com/2016/03/22/james-hansens-apocalyptic-sea-level-study-lands-to-mixed-reviews/

Downs, A. 1972. Up and Down with Ecology: The Issue Attention Cycle. *Public Interest* 28(1): 38–50.

Edwards, B., and J. D. McCarthy. 2004. Resources and Social Movement Mobilization. In *The Blackwell Companion to Social Movements*, eds. D. A. Snow, S. A. Soule, and H. Kriesi, 116–152. Blackwell.

Endres, D., L. M. Sprain, and T. R. Peterson. 2009. *Social Movement to Address Climate Change: Local Steps for Global Action*. Cambria Press.

Ganz, M. 2004. Why David Sometimes Wins: Strategic Capacity in Social Movements. In *The Psychology of Leadership: New Perspectives and Research*, eds. D. M. Messick and R. M. Kramer, 209–240. Psychology Press.

Gitlin, T. 1980. *The Whole World is Watching*. University of California Press.

Giugni, M. 2004 *Social Protest and Policy Change: Ecology, Antinuclear, and Peace Movements in Comparative Perspective*. Rowman and Littlefield.

Han, H., and N. Stenhouse. 2015. Bridging the Research-Practice Gap in Climate Communication: Lessons from One Academic-Practitioner Collaboration. *Science Communication* 37(3): 396–404.

Hansen, J., M. Sato, P. Hearty, R. Ruedy, M. Kelley, V. Masson-Delmotte, G. Russell, et al. 2016. Ice Melt, Sea Level Rise and Superstorms: Evidence from Paleoclimate Data, Climate Modeling, and Modern Observations that 2 °C Global Warming Could be Dangerous. *Atmospheric Chemistry and Physics* 16(6): 3761–3812.

Hestres, L.E. 2015. Climate Change Advocacy Online: Theories of Change, Target Audiences, and Online Strategy. *Environmental Politics* 24(2): 193–211.

Kitschelt, H.P. 1986. Political Opportunity Structures and Political Protest: Anti-Nuclear Movements in Four Democracies. *British Journal of Political Science* 16(1): 57–85.

Leiserowitz, A., E. Maibach, C. Roser-Renouf, and J. D. Hmielowski. 2012. Extreme Weather, Climate and Preparedness in the American Mind. Yale Project on Climate Change Communication. http://environment.yale.edu/climate/files/Extreme-Weather-Climate-Preparedness.pdf

Maney, G. M., R. V. Kutz-Flamenbaum, D. A. Rohlinger, and J. Goodwin, eds. 2012. Introduction. Strategies for Social Change (Social Movements, Protest and Contention. University of Minnesota Press.

McCarthy, J.D., and M.N. Zald. 1977. Resource Mobilization and Social Movements: A Partial Theory. *American Journal of Sociology* 82(6): 1212–1241.

McComas, K., and J. Shanahan. 1999. Telling Stories About Global Climate Change: Measuring the Impact of Narratives on Issue Cycles. *Communication Research* 26(1): 30–57.

McCombs, M.E., and D.L. Shaw. 1972. The Agenda-Setting Function of Mass Media. *Public Opinion Quarterly* 36(2): 176–187.

Meyer, D.S., and D.C. Minkoff. 2004. Conceptualizing Political Opportunity. *Social Forces* 82(4): 1457–1492.

Mooney, C. 2011. Climate-Media Paradox: More Coverage, Stalled Progress. http://www.desmogblog.com/climate-media-paradox-more-coverage-stalled-progress

Nisbet, M.C. 2014. Disruptive Ideas: Public Intellectuals and Their Arguments for Action on Climate Change. *Wiley Interdisciplinary Reviews: Climate Change* 5(6): 809–823.

Oliver, P.E., and G. Marwell. 1988. The Paradox of Group Size in Collective Action: A Theory of the Critical Mass II. *American Sociological Review* 53(1): 1–8.

Olzak, S., and E. Ryo. 2007. Organizational Diversity, Vitality and Outcomes in the Civil Rights Movement. *Social Forces* 85(4): 1561–1591.

Palmer, L. 2012. Whose Is the Face, and the Voice, of Climate Change? Yale Forum on Climate Change and The Media. http://www.yaleclimatemediaforum.org/2012/03/whose-is-the-face-and-the-voice-of-climate-change/

Pearson, A.R., and J.P. Schuldt. 2014. Facing the Diversity Crisis in Climate Science. *Nature Climate Change* 4(12): 1039–1042.

Polletta, F., and J.M. Jasper. 2001. Collective Identity and Social Movements. *Annual Review of Sociology* 27: 283–305.

Priest, S., T. Greenhalgh, H.R. Neill, and G.R. Young. 2015. Rethinking Diffusion Theory in an Applied Context: Role of Environmental Values in Adoption of Home Energy Conservation. *Applied Environmental Communication & Education* 14(4): 213–222.

Roberts, D. 2011. 'Brutal Logic' and Climate Communications. *Grist*. http://grist.org/climate-change/2011-12-16-brutal-logic-and-climate-communications/

Romm, J. 2011. What Mistakes Did the Environmental Community and Progressive Politicians Make in the Climate Bill Fight. *Climateprogress*. http://thinkprogress.org/romm/2011/04/23/207955/what-mistakes-did-the-environmental-community-and-progressive-politicians-make-in-the-climate-bill-fight/

Saad, L. 2015. Americans Choose "Pro-Choice" for First Time in Seven Years. http://www.gallup.com/poll/183434/americans-choose-pro-choice-first-time-seven-years.aspx

Skocpol, T. 2013. Naming the Problem: What It Will Take to Counter Extremism and Engage Americans in the Fight Against Global Warming. http://www.scholarsstrategynetwork.org/sites/default/files/skocpol_captrade_report_january_2013_0.pdf

Skocpol, T., and V. Williamson. 2011. *The Tea Party and the Remaking of Republican Conservatism*. Oxford University Press.

Smil, V. 2010. *Energy Transitions: History, Requirements, Prospects*. Praeger.

Sprain, L. 2015. Framing Science for Democratic Engagement. Unpublished paper, University of Colorado Boulder.

Sutter, J. 2016. Maybe Stop Selling the Ocean? http://www.cnn.com/2016/03/24/opinions/sutter-new-orleans-climate-auction/

Sweetser, K.D., G.J. Golan, and W. Wanta. 2008. Intermedia Agenda Setting in Television, Advertising and Blogs During the 2004 Election. *Mass Communication and Society* 11: 197–216.

Swift, A. 2015. America's Desire for Stricter Gun Laws Up Sharply. http://www.gallup.com/poll/186236/americans-desire-stricter-gun-laws-sharply.aspx

Walsh, B. 2011. The Unfair Reception of the Climate Shift Report Shows that Greens Need to be More Open to New Ideas. *Time*, April 25. http://ecocentric.blogs.time.com/2011/04/25/battling-over-the-climate-war/

Wynn, G. 2012. Climate Science Uncertainty Impacts Discourse. *Huffington Post*, January 26. http://www.huffingtonpost.com/2012/01/26/climate-science-uncertainty-effects_n_1233244.html

CHAPTER 8

The Path Forward: Making Change Happen

To understand how to develop and study (as scholars) or implement (as communicators) effective communication strategies in a climate change context will require something of a paradigm shift in our thinking. Science communication scholarship and practice have already changed in major ways. As a community, we are well beyond the era where we could assume that the dissemination of accurate scientific information would be a sufficient antidote to low levels of scientific literacy or solve the "problem" of science-society relationships. Dialogue, discussion, and other forms of two-way participation involving non-scientists in reflecting on scientific developments (and in some cases, even contributing to those developments) are more likely to be promoted by today's science communication scholars and progressive practitioners, but they have their limits as well. These events and activities generally take place on a small scale, are often one-time experiments, and do not readily "scale up" to facilitate broader participation. Those who do participate may tend to be those already actively interested in scientific and technological ideas. And the link between these events and activities and collective action on policy is weak, both conceptually and in practice.

Many of our existing communication models and persuasive strategies have been developed in quite different contexts, such as health communication or political communication or product advertising or the diffusion of innovations. They do not always translate to climate in a simple way.

© The Author(s) 2016
S. Priest, *Communicating Climate Change*,
Palgrave Studies in Media and Environmental Communication,
DOI 10.1057/978-1-137-58579-0_8

They also do not translate well to understanding and facilitating collective action. In this book, we have emphasized that collectivities matter, whether we are talking about professional norms and ethics in journalism and science, the social networks that underlie social media, the social organization of science, or—in the end—the kind of broad-based social movement that will be needed to combat climate change in our century (we hope sooner rather than later). Our research paradigm needs to shift to a clearer focus on the collective. We need new theoretical work here, as well as new research strategies. We also need to rethink how we are using familiar concepts such as efficacy, emotional appeals, science literacy, and group identity in the context of promoting social change. To help people understand the basics of the science underlying climate change is one thing, but to help them understand what we need to do about it is quite another.

Our traditional mass media are far from perfect, but they have made progress on climate reporting. Will this be enough? With fewer journalistic gatekeepers and vastly more sources to choose from, thanks to rapid technological development, individual citizens are empowered to make new choices among a far larger number of competing voices, but they may need new skills to do so effectively. Among these skills is an understanding of what it means to suggest that science itself is a collective social process and that recognizing uncertainty is a part of that process and not always a sign of mistakes being made—or that scientific knowledge cannot be trusted. Meanwhile, journalism itself has largely been redefined. The old problem of "lopsided" reporting through false balance may have receded, but is itself a part of the bigger problem of indiscriminately conveying scientific legitimacy (or, sometimes, taking it away). But that bigger problem is now resting more squarely on the shoulders of individual information consumers who must navigate through a new landscape to arrive at opinions on complex topics—or else simply take the word of leaders they respect. Few scientists are among those leaders, and even fewer are both good scientists and good communicators. The trusted news anchor is a vanishing species. We also need new research on how opinion leadership and credibility operate, for climate issues, in the world of new media, as well as studies of other dynamics of today's opinion climate for science.

The news media as a whole may still set the agenda, but the media landscape and the social ecology in which this landscape is embedded are far more complex now than they seemed a generation ago. Blogs, informational websites, social media messages, partisan cable news networks, startup online news organizations, and myriad other influences shape our perceptions of the world in new and seemingly more individualized ways. Traditional media such as newspapers, magazines, and network television news continue to be influential, but they have more competition for people's attention and trust; print media in particular have been through many years of declining presence and influence. The newer media may serve to isolate us as much as they serve to unite us and to overload us as much as to inform us. How do we best reach people and change minds in this new media world? What is the available path forward toward collective action on climate change?

In the end, addressing climate change will require much more than a better-informed citizenry. Individual lifestyle changes are important but not enough. Ultimately, success will require the creation of new paths to action, beginning most logically at the local level where the impacts of climate change are already being felt. Ultimately, ongoing action at the federal and even the global level are also completely essential, but this seems an even greater challenge. Climate has no clearly defined constituency; it will affect—indeed, it is already affecting—everyone, and yet even those most likely to be affected in the near term have not mobilized to advocate for available solutions. Many may not recognize how their local weather problems connect to global climate trends. Environmental groups will be part of the necessary mobilization, but—like politicians—they have their own constituencies and their own priorities. Climate plays an important part in the work of many such groups, but we need to make climate more central to *someone* in order to educate and engage *everyone*, as well as to put new policies in place.

How can social action coalesce around an issue so diffuse, so seemingly distant in time and space and so technically complex and yet so threatening that it seems some people simply opt out of thinking about it at all? What is it about our most cherished values, our deepest collective identity, that should be motivating us to move forward on this front and yet, at the same time, seems to hold many of us back? Social movements may attract members because those members feel a sense of belonging and commitment, but also because movement goals and identities reinforce the particular kind of collective identity that the movement stands for. What does climate stand for, then, in this sense? Communication scholarship should be able to address this and related issues, but it will take some rethinking and retooling. Instead of continuing to search for a messaging frame that will overcome all of this inertia, we need to ask some new questions that will likely require new approaches.

Climate is a social justice issue. All people on the globe will not be affected equally, and some are in a better position to shelter themselves against the worst effects of new climate patterns than others. Future generations, rather than those alive today, will be more affected than present ones. Older people are more susceptible than others to heat stress

(U.S. EPA 2016). Poorer people worldwide have fewer resources with which to cope with changes. And coast dwellers in all locations (rich, poor, rural or urban) face particular challenges. A very good argument can be made that we—that is, those among the people of the present who have the resources to support action—have an ethical obligation to address this problem on behalf of others, including future generations, on top of the many short-term, practical, more self-interested reasons to do so. And in the end, concern about climate likely needs to coalesce into its own social movement, given that the present array of environmental and conservationist NGOs does not seem well positioned to push forward in concert on this issue. Neither scientists nor environmentalists can or should shoulder the entire burden of advocating for reasoned climate policy.

Communicators and communication scholars cannot shoulder it either, but the communication community might well take note of the research opportunities provided—and of the fact that answers to new research questions about the relationship between communication and social action should prove useful to climate advocates. This movement may not look a lot like previous U.S. social movements for civil rights or environmental protection; that remains to be seen. The problem of climate will take us in new directions in this new media, post-mass communication world. Even so, we can learn a lot from the literature on previous social movements that is applicable here, as Chap. 7 demonstrated.

We have tried to emphasize throughout this book, without discounting the need to develop better ways to communicate and persuade on an individual basis, that our field should not overlook consideration of broader collective processes—or the research opportunities they provide. Human beings are social, communication is critical to that characteristic, and climate change needs collective social action. What in fact glues us together as societies (local, national, global) is in large part communication—one of the most notable and truly remarkable hallmarks of ourselves as a species. Public opinion is measured at the individual level, but it is formed and altered through collective processes, and its collective dynamics influence us on both the individual and the collective level. Communication media remain a central factor, even if far more dispersed and diverse than ever before, and influence the perceived opinion climate in new ways. We need much more research on how these processes of change play out.

In light of the above discussion, we would like to suggest four specific directions forward for communication specialists and others interested in improving communication about climate: continue talking to people, face-to-face and via mediated communication, about climate—so it stays on the public agenda; consider a "turn to the collective" in our research, further emphasizing the relationship between communication and collective processes; understand climate change as a social justice issue that requires a social action solution, not just behavior change at the individual level; and facilitate that action by a focus on presenting solutions—not just problems.

KEEP TALKING! INTERPERSONAL STRATEGIES MATTER

We have already mentioned that when Hurricane Katrina rolled over New Orleans in 2005, some people left their homes earlier than others. Our work, based on interviews with Katrina evacuees across the U.S. South (Taylor et al. 2009), asked 114 people what went through their minds at the moment of their final decision to leave. When they focused on remembering that moment, many of them reported that they already knew about the storm and understood its likely severity via some combination of media reports, interpersonal sources, and previous personal experience, but it often took a personal message—a neighbor knocking on the door, an urgent telephone call, in a few cases even a rescue worker—to push them to the point of an actual evacuation decision with the message that "it's time to go". Storm information by itself, in other words, was not enough. Less than a third of those we talked to left because of media reports alone, although remarks by trusted media representatives and familiar political voices on broadcast news were sometimes persuasive. In many cases it also took an evolving *collective* response resulting in a person-to-person message diffusing through a local social network to elicit action.

Communication scholars have known for decades that interpersonal communication or a combination of interpersonal and mediated communication can be much more powerful than mediated communication alone (see Lazarsfeld et al. 1944). Interpersonal communication travels rapidly through social networks; this is not just a matter of one person speaking to one other person, a fact made more obvious in the Internet age. We also know that trust is an extremely important factor in risk communication, even if this means trusting a source seen on television—which may be experienced as a close parallel to trusting a physically present person. Face-to-face and mediated communication coming from trusted spokespersons will continue to be a vital component of climate communication going forward. Yet many people do not know any scientists personally, and scientists seem to speak more often through print media and blog texts than in broadcast, podcast, or video formats. This underscores the need for others to fill the "opinion leader" role, ranging from outreach and education staff at parks, museums and universities to teachers, writers, and media figures, political and religious leaders, and perhaps most importantly relatives, neighbors and friends. Even communication scholars can fill this role in their private lives, as can natural scientists.

Speaking out about climate matters. Not only does it keep the issue on the media and therefore the public agenda, but it shapes the climate of public opinion and nurtures the expansion of people's background knowledge. This process transfers to the "new media" world, which multiplies person-to-person outreach in new ways. Businesses are typically very pleased when their messages "go viral", one of the least expensive advertising strategies ever imagined since after the original production and placement (which may be low cost itself), the diffusion of the message is completely free. Yet the key characteristics that make this happen for one message and not another remain unknown—another research opportunity for communication scholars. This is also more than one person "talking" (via social media) to another; it is multiplied by the connection between technological networks and true social ones. For those looking for new research opportunities, we still have much to learn about both kinds of networks and how messages pass between and through them.

However, actual in-person, face-to-face communication with other individual people, even in casual settings, matters too—it may be even more important, since Internet messages are easily ignored. Every time someone brings up the weather in a conversation, for example, it can be thought of as a teachable moment. Communication researchers should consider pursuing research questions that focus on these issues as well, perhaps through a fieldwork, interview, or focus group effort.

Focus on the Collective: A Renewed Research Paradigm

Communication scholars Boudet and Bell (2015), writing about the nature of social movements, argue that risk communication work has not focused enough on the role of social groups, as opposed to individuals. We very strongly agree. Why would this be the case? Sociology, the study of human social life, was an early foundation of communication scholarship, mass communication scholarship in particular. Although there were multiple other influences, many early communication scholars who later became well known had sociology training. Yet somewhere along the line, our research tended to become rather more reductionist. Group membership is routinely reduced to demographic variables in quantitative studies in which the individual is most often the unit of analysis. That often reflects a missed opportunity to better understand just how the group influences the individual—and conversely, how individuals (especially those in opinion leadership positions) affect the group.

Was this turn toward quantitative work a function of communication scholarship seeking its place in the academic world? Sophisticated quantitative studies impress tenure committees and help cement the argument that communication is a rigorous field. That is not at all to say that experimental or survey work is not valuable or that questions of academic status are the only factors that motivate it. On the contrary, this work absolutely continues to give us important new insights, but these approaches do typically tend to reinforce a focus on the individual rather than the social dynamics that are in play. In general, if our goal is to understand social action decisions and the dynamics of opinion formation, we will also need more studies based on ethnographic, case study, or interview-based approaches, as well as more true mixed-methods research. Although the quantitative-qualitative divide in social science research is sometimes a touchy issue, it should not be. Mixed-methods efforts might be particularly suited to understanding how advocacy and movement organizations form and function.

Early sociology and the communication work that began there reflected creativity that sometimes seems missing from today's communication scholarship. Asking theoretical questions that are vitally important for human society might help bring that back; climate change provides an opportunity to do just that. Communication research needs to breathe new life into studies of collective phenomena. As social scientists, we should take better account of the central role within social groups played by communication. Communication cannot take place absent a social context and a shared language and culture. This does not just involve information transfer, of course, but things like network construction, group decision making, and identity formation. As a scholarly community, we would benefit from further diversifying our research approaches.

For the present discussion, perhaps the most important place where this issue arises is in considering the kinds of values and identities that will make a climate movement happen. Here, the individual intersects with the social. However, we should not assume that we can always adequately measure things like collective identity or the dynamics of collective behavior with experimental or survey-based approaches alone. It is not yet clear what kind of collective identity would prompt people to join a climate movement, representing another research opportunity—one where well-designed experiments might contribute greatly. But, while we may eventually find quantitative variables that can "stand in" for some of the relevant social elements, these variables usually do not tell the whole story. Human social identity and human culture are sometimes better understood holistically and not as answers to a discrete series of questions. And we are looking at a period of exploratory work before we find some of our answers.

Action Orientation: Climate as a Social Justice Issue

If we look at successful social movements that have taken place in the United States, they share a single characteristic that may easily be overlooked: an orientation to social justice and to the inalienable rights of human beings. Think about the abolition of slavery, the women's suffrage movement, the later feminist movement, and a whole series of somewhat interlocked civil rights movements surrounding ethnicity, not only for African Americans, but for Asian Americans, Hispanic and Latino Americans, and Native Americans, as well as senior citizens, people with disabilities, and most recently LGBT (lesbian, gay, bisexual, and transgender) people. The anti-Vietnam War movement ultimately changed the course of U.S. policy in Southeast Asia—and, arguably, ended or at least helped to end the war itself. There, both the rights of the Vietnamese to self-determination and the rights of young Americans to resist a military draft forcing them to fight in what was widely considered an unjust war were at stake. This history suggests that arguments about the rights of future generations and of contemporary people in the areas of the world most vulnerable to climate change should also prove effective.

Worth special mention in this context as a strong example of changing attitudes culminating in political action: In 1988, President Ronald Reagan signed a bill to award $20,000 and a formal apology from the government to each of well over 100,000 Japanese Americans (a majority of them U.S. citizens) who were sent to prison camps on U.S. soil during World War II (Qureshi 2013). Admittedly, this small an amount of money could never compensate those victims for their experience, on the one hand, but on the other, this demonstrates that America does have at least something of a collective historical conscience where human rights are concerned.

At first glance, the environmental movement may not seem to fall into the same pattern of human rights orientation, but actually it can be interpreted as representing a movement reaction on behalf of the rights of ordinary citizens to live and work in an unpolluted environment that is not dangerous to them or to other species they care about. Some (although certainly not all) environmentalists extend the right to a continued and stable existence to many other species. Protests have reduced seal kills, led to increased protection for whales, dolphins, and other sea life, and stopped the logging of old-growth forests (primarily in cases where critical animal habitat has been involved, e.g., that of the spotted owl in the U.S. Pacific Northwest). Public pressure has resulted in the re-introduction of wolves in Yellowstone National Park, one of the most-loved parks in America. Climate will affect many remaining wild species (as well as domesticated ones), so advocates for those species and their organizations should be ready recruits to the climate change cause—even though they may not always be its leaders.

Communication scholars who are particularly attracted to studies of collectivities should consider action research as an alternative form of scholarship (see Abraham and Purkayastha 2012, for a review introducing a special issue of *Current Sociology* on this topic). In a departure from the usual assumption that researchers must remain psychologically separated from those they study in order to be objective, action research assumes instead that researchers can be an embedded component of an active organization or community that is trying to solve problems and pursue new directions. While many action research projects are carried out in the developing world, where visiting researchers may help local communities with limited resources to define, articulate, and then realize their own goals, communication scholar Lana Rakow (2005) reports on her own action research approach to study of her local community after a major flood. She calls the ongoing failure of communication researchers to study communities (including their own) "disappointing and troubling" (p. 6).

Communication researchers who have the time, interest, commitment and resources to involve themselves in action research within climate movement organizations, emerging alternative energy entrepreneurs, or other social contexts where people's work and lives intersect centrally with climate issues face a rewarding opportunity. This work could potentially target communities affected by climate-related changes in patterns of flooding and drought or resource availability issues (such as reductions in fish populations and forests lost to wildfire), including Native American communities who may be among the especially vulnerable (U.S. EPA 2016). These researchers could offer communication-related advice and assistance to those organizations or communities.

Push Out Solutions, Not Just Problems

Climate likely seems to some a problem that not only has primarily remote effects but also has elusive and remote solutions. Although the role of efficacy (a sense of control through action) in health communication and other situations is well established, it has not been as firmly established for climate change. Perhaps this is because no matter the wording of a survey question or the experimental message condition being tested, research participants simply do not feel efficacious with respect to climate. Of course, as researchers, we need to look at this question more deeply. But meanwhile, the role of efficacy being so well-established in many other arenas, practitioners should take seriously the idea that people need to know what it is they can in fact do—turn off the lights, turn down the thermostat, and travel less, yes. Many other things could be added to that list. But there must also be advocacy and support for new legislation.

Necessary solutions will not only go beyond lifestyle adjustments, but also beyond the establishment of controls on carbon and other greenhouse gas emissions or to the preservation of tropical forests. We need to pro-actively extend investment in alternative energy and continue to provide positive incentives for its adoption and use. Research designed to clear a path for change needs to recognize the collective nature of political will and in some cases it may make sense to actively facilitate action on climate issues from an action research perspective. To create more activism around these issues, we will also need to demonstrate hope that such efforts can be successful. And to create hope, knowing even small steps people can take will help enormously—not just to encourage action, but to alleviate a sense of helplessness. Together, we can make progress.

References

Abraham, M., and B. Purkayastha. 2012. Making a Difference: Linking Research and Action in Practice, Pedagogy, and Policy for Social Justice: Introduction. *Current Sociology* 60(2): 123–141.

Boudet, H. S., and S. E. Bell. 2015. Social Movements and Risk Communication. In *The Sage Handbook of Risk Communication*, eds. H. Cho, T. Reimer, and K. A. McComas, 304–316. Sage.

Lazarsfeld, P. F., B. Berelson, and H. Gaudet. 1944. *The People's Choice: How the Voter Makes Up His Mind in a Presidential Campaign*. Columbia University Press.

Qureshi, B. 2013. From Wrong to Right: A U.S. Apology for Japanese Internment. http://www.npr.org/sections/codeswitch/2013/08/09/210138278/japanese-internment-redress

Rakow, L. 2005. Why Did the Scholar Cross the Road? Community Action Research and the Citizen-Scholar. In *Communication Impact: Designing Research that Matters*, ed. S. Priest, 5–18. Rowman & Littlefield.

Taylor, K., S. Priest, H.F. Sisco, S. Banning, and K. Campbell. 2009. Reading Hurricane Katrina: Information Sources and Decision-Making in Response to a Natural Disaster. *Social Epistemology* 23(3): 361–380.

U.S. Environmental Protection Agency. 2016. Climate Impacts on Society. https://www3.epa.gov/climatechange/impacts/society.html

INDEX

A
action orientation, 168–70
action research, 169, 170
activism, 110
agenda-building, 69
agenda-setting, 69, 138
alliances, 149
alternative energy, 170
American Association for the Advancement of Science, 68, 90
anti-nuclear movement, 149, 150
anti-Vietnam War movement, 168
astrology, 129
attention cycle, 137, 140. *See also* issue-attention cycle
attentive public, 47
audiences, 10, 27, 43–6
autism, 123

B
background knowledge, 154
balance, 29, 76, 99, 126
best practices, 154
broadcast news, 96

C
Centers for Disease Control and Prevention, 107
citizen deliberation, 91
citizens, 116, 121, 128–30
citizen science, 8, 128
civil rights movements, 146, 147, 168
climate, 41
  action, 111
  movement, 144, 148, 154. *See also* social movements
  of opinion, 14, 36, 39–41
  of professional opinion, 71
  of public opinion, 13, 33, 45, 65, 78, 117
climategate, 46
cognitive dissonance, 26
cold fusion, 100, 101, 123, 125
collective, 34, 110, 116, 119
  action, 32, 39–41, 49, 57, 152, 161–3
  behavior, 12, 167
  identity, 147–9, 153, 163, 167
  level, 2

memory, 138
processes, 66, 72, 121, 164
communication research, 1, 11
communities, 169
confirmation bias, 110
consensus, 74, 77, 120, 122
contemporary developments, 89
contrarian views, 127
COP 21, 58, 92, 141
credentials, 120
credibility, 162
critical science literacy, 38, 120, 121, 124, 126, 127, 130, 132. *See also* science literacy
cues, 116. *See also* heuristic
cultivation, 34, 35

D
deficit, 91, 103, 152
model, 7–9, 103, 130
"deficit" to "dialogue", 91, 115, 117
deliberative democracy, 93
democracy, 117, 124, 126, 128, 131–2
democratic governance, 15
demographic factors, 52
"denier" perspectives, 4
Department of Defense, 106
developing world, 169
dialogue, 7, 91, 161. *See also* "deficit" to "dialogue"
diffusion theory, 110
dissent, 126, 127
distribution of power, 25

E
earthquake, 125
efficacy, 54, 57, 109, 110, 162, 170
emotional appeals, 162
energy use, 27, 48
engaging the public, 130, 131

environmental activist, 143
Environmental Defense Fund (EDF), 104, 105
environmental groups, 106
environmentalist community, 105
environmentalist values, 110
environmental journalists, 78
environmental movement, 40, 169
environmental NGOs, 106
environmental organizations, 103–5, 145, 148
ethical norms, 73, 75, 82, 83
ethics of communication, 81
ethos, 121
evolution, 24

F
face-to-face, 166
false balance, 13, 29, 77, 137
fatalistic beliefs, 53
fear, 47, 109, 110
fracking, 105
framing, 15
free speech, 70
functionalism, 25
Futurity, 101

G
gatekeeping, 97, 101, 138
government agencies, 67, 68
greenwashing, 98
group identity, 162

H
heuristic, 109, 116
cue, 55
processing, 109, 115, 118
Heuristic-Systematic Processing Model, 108
hierarchy of influence, 78

home energy conservation, 110
hope, 110
how science works, 117
Hurricane Katrina, 37

I
ideologies, 53, 116
individualistic, 66
information seeking, 44, 84, 109, 110
information subsidy, 29, 90, 98–100, 102, 138, 144
information system, 28–30
Intergovernmental Panel on Climate Change (IPCC), 69, 140
inter-media agenda setting, 141
Internet, 95, 108, 124, 126
interpersonal, 165–6
  communication, 41, 55
  conversation, 34
issue-attention cycle, 138, 141, 142

J
Japanese Americans, 168
journalism, 16, 96
journalistic ethics, 75
journalistic "objectivity," 35
journalistic routines, 76
journalists, 17, 71, 89, 124
judgmental discounting, 56

K
knowledge brokers, 95, 97, 99, 102

L
legitimacy, 34, 40, 79, 97, 123, 124, 126, 137, 142, 144
linear transmission, 101
lobbying, 150
local grassroots movements, 154

M
making sense of science, 122
mass communication, 41, 43
mass media, 44, 162
media attention, 144
media legitimization, 79
medialization, 71, 101
medialized, 77
media representations, 34
media system, 27
Merton, Robert, 121
Mertonian norms, 80
mixed-methods research, 167
mobilization, 37, 143, 146, 150, 153
motivated reasoning, 110
movement organizations, 148, 152, 167. *See also* social movements

N
National Academy of Sciences (NAS), 68
National Association of Science Writers, 78
National Citizens' Technology Forum, 91
National Science Foundation, 82
Native American, 169
networks, 33
New Madrid earthquake, 123
new media, 121, 124, 162, 166
nonprofit organizations, 69
normal science, 122
norms, 74
nuclear power, 105

O
objectivity, 76, 99, 169
one-way transmission model, 80
opinion formation, 167

opinion leaders, 11, 130, 162, 166, 167
ozone depletion, 48

P
paradigm shift, 161
participation, 153
partisan gap, 52
peer review, 121
personal observations, 48, 53
persuasion, 16, 27, 32, 37, 41, 55
pluralistic, 117, 133
polar bears, 104
polarization, 94, 133
policy change, 46
political activism, 110
political affiliation, 52
political agenda, 144
political opportunity structure, 149, 150, 154
political party, 52
political will, 170
population control, 104
positive action, 109
presidential election, 133
professional associations, 70–3
professional norms, 72
pseudo-scientific, 120
psychometric paradigm, 50
public, 44, 45
   engagement, 7, 83, 90, 91, 132
   health agencies, 68
   opinion, 3–5, 28, 49
   participation, 94
   relations, 76

R
rational, 117, 120
reductionist paradigm, 33
religion, 54

religious, 53
research paradigm, 166–8
resource mobilization, 145, 146, 152
responsibility, 57
risk, 51
   communication, 54, 166
   information seeking, 45
Risk Information Seeking and Processing Model (RISP), 107, 108, 110

S
science blogs, 82
science cafés, 8
science communication, 4
science journalism, 107, 125, 131
science literacy, 16, 48, 115, 118, 120, 122, 125, 128, 129, 132, 161, 162. *See also* critical science literacy
   measurement of, 129
scientific consensus, 4, 17, 36, 77, 79, 122
scientific dissent, 76
scientific journals, 68
scientific method, 17
scientists, 37, 38, 71–3, 80, 81, 90, 99
setting the agenda, 137
Sierra Club, 104–6
single-study science stories, 36
"skepticism", 25–9
slow-moving disaster, 144–5
smoking, 127
social, 119–24
   action, 163–5, 167
   capital, 146
   change, 25, 40
   context, 137
   ecology, 94, 97, 162
   enterprise, 132
   institutions, 131

justice, 148, 163, 165, 168–70
media, 56, 82, 91
movements, 5, 14, 30, 40, 137, 142–5, 147, 151–3, 163, 164. *See also* climate movement
networks, 27, 31, 37, 41
norms, 73–5
process, 118, 131, 162
socially amplified, 51
Society of Professional Journalists, 76
sources, 72, 76, 127
spiral of silence, 11, 65, 117, 148
stem cell, 123, 125, 126
stewardship, 54
strategic capacity, 151–2, 154, 155
strategic communication, 15, 16, 45, 66
strategic planning, 151
sustainable, 149
systematic processing, 109, 115, 116
system justifiers, 53

T
targeted marketing, 44
technological fix, 27
theory of planned behavior, 108, 110

threat appraisals, 51
Three Mile Island, 150
transmission model, 99
trust, 39, 54–6, 116, 120–2, 125, 165
  in experts, 55
two-way communication, 9

U
uncertainty, 35, 49, 52, 77, 79, 118, 126, 127, 140
unit of analysis, 9–15

V
values, 47, 50, 116, 119, 163
visible scientists, 83
volunteers, 71

W
weather, 48, 77
wind energy, 106
weathercasters, 56
weight-of-evidence, 78
World Wide Views, 92, 93

Printed in the United States
By Bookmasters